Primary Maths for Scotland

2nd Level Maths
Textbook 2B

Series Editor: Craig Lowther

Authors: Antoinette Irwin, Carol Lyon,
Kirsten Mackay, Felicity Martin, Scott Morrow

Contents

1 Estimation and rounding

1.1 Rounding whole numbers to the nearest 10, 100 or 1000

We are learning to round numbers to the nearest 10, 100 or 1000.

Before we start

Choose five numeral cards from a set of numeral cards from 0 to 9.
Arrange the digits to make the smallest number you can from the cards you picked.
What number have you made?
What value does each digit represent?

We can use a number line to decide whether to round up or round down.

Let's learn

To round 3287 to the nearest 10, we look at multiples of 10 on either side. 3287 is between 3280 and 3290, but it is closer to the number 3290 so we round up to 3290.

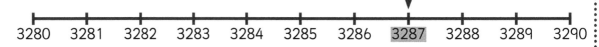

3280 3281 3282 3283 3284 3285 3286 3287 3288 3289 3290

To round 3287 to the nearest 100, we look at multiples of 100 on either side. 3287 is between 3200 and 3300 but it is closer to the number 3300 so we round up to 3300.

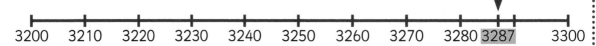

3200 3210 3220 3230 3240 3250 3260 3270 3280 3287 3300

To round 3287 to the nearest 1000, we look at multiples of 1000 on either side. 3287 is between 3000 and 4000 but it is closer to the number 3000 so we round down to 3000.

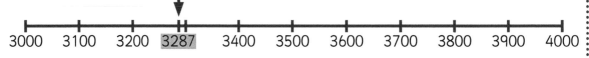

3000 3100 3200 3287 3400 3500 3600 3700 3800 3900 4000

If the number ends in 5, 50, 500 or higher, we usually round up.

1) Write the two multiples of 100 that are either side of the number, then round to the nearest 100. Draw a number line to help you.

   a) 632 lies between   and   , rounded

   b) 7869 lies between   and   , rounded

2) Has the newspaper headline been rounded correctly to the nearest 1000?

   a) **77 000 attend football match!**   Actual attendance 76 575
   Yes / No

   b) **Four thousand visit exhibition.**   Actual number of visitors
   3698 **Yes / No**

   c) **£5000 raised for charity**   Actual amount raised
   £4499 **Yes / No**

   d) **3000 students attend Maths lecture**   Actual number at the lecture
   2611 **Yes / No**

3) Visitor numbers at the museum were 4568 for March, 5429 for April and 7347 for May. Draw a table to record each of the months rounded to the nearest 10, nearest 100 and nearest 1000.

CHALLENGE!

Work with a partner. You will need a set of numeral cards from 0 to 9. Shuffle the cards and lay them out face down. Take it in turns to pick four numeral cards and make a four-digit number of your choice. Ask your partner to round the number to the nearest 10, 100 and 1000 for the number you chose. Check their answers are correct! Then swap roles.

Now try picking five cards to make a five-digit number, then six to make a six-digit number.

1 Estimation and rounding

1.2 Rounding decimal fractions to two places

We are learning to round decimal fractions to the nearest tenth or whole number.

Before we start

What does the digit 5 represent in this number? How much is the digit 6? What would the number one tenth more than this number be? How about one hundredth more?

2·56

We can round decimal fractions up or down to the nearest tenth or whole number.

Let's learn

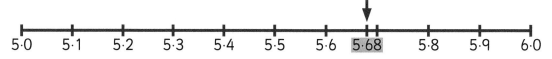

5·0 5·1 5·2 5·3 5·4 5·5 5·6 5·68 5·8 5·9 6·0

5·68 is between 5 and 6, but is closer to 6, so we round up to 6.

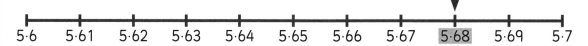

5·6 5·61 5·62 5·63 5·64 5·65 5·66 5·67 5·68 5·69 5·7

5·68 is between 5·6 and 5·7, but is closer to 5·7, so we round up to 5·7.

When we round decimal fractions to the nearest whole number, we look at the first digit after the decimal point.

When we round decimal fractions to the nearest tenth, we look at the second digit after the decimal point.

If this digit is 5 or more, we usually round up to the nearest whole number.

If this digit is less than 5, we usually round down to the nearest whole number.

Let's practise

1) Fill in the missing numbers:

a) 6·89 rounded to the nearest whole number is _____.

b) 0·45 rounded to the nearest tenth is _____.

c) 1·89 rounded to the nearest tenth is _____ and the nearest whole number is _____.

2) Can you round these numbers to the nearest tenth and the nearest whole number?

	Nearest tenth	Nearest whole number
5·56		
19·26		
17·01		
206·69		
1·48		

3) Are these number sentences true or false? Circle true or false.

1·31 rounded to the nearest tenth is 1·4.   **true/false**

85·55 rounded to the nearest tenth is 85·5.   **true/false**

10·04 rounded to the nearest tenth is 10·4.   **true/false**

3·71 rounded to the nearest whole number is 4.   **true/false**

19·78 rounded to the nearest whole number is 19·8.   **true/false**

CHALLENGE!

Sarah is shopping. She has a 10 pound note to spend, but isn't sure if she has enough money to buy all these items:

Packet of pens   £1·32   Notepad   £4·15
Sharpener   £0·85   Highlighters   £2·97

Can you use rounding to estimate if she has enough money? Explain your answer.

1 Estimation and rounding

1.3 Using rounding to estimate the accuracy of a calculation

We are learning to use rounding to estimate and check our answers to calculations.

Before we start

Can you round this number to the nearest 10? Nearest 100? Nearest 1000?

24 569

Rounding numbers makes it easier to estimate the answer to a calculation.

Let's learn

When you are solving problems, it is always a good idea to check your answer to see if it is a reasonable answer.

Nuria and Isla work out the problem 762 – 419. Nuria works out the answer as 271 and Isla gets 343.

To see who is right we can estimate the answer by rounding to the nearest 10:

760 – 420 = 340

We can see that Nuria's answer is clearly not reasonable, but Isla's answer is much closer to our estimation, so she is more likely to be correct.

Let's practise

1) Finlay got his bill at a restaurant and decided that it was incorrect.

   Use rounding to estimate the total of the bill. Do you agree with Finlay? Explain why.

Bridge Street Restaurant		
Roast beef	1	£13·50
Fish & Chips	2	£23·00
£11.50 × 2		
Steak & Chips	1	£18·50
Drinks	1	£14·00
******************************		
TOTAL		£99·00

2) Estimate whether each answer is reasonable or not and explain your answer.
   a) 2341 + 5718 = 7059
   b) 1699 – 324 = 1375
   c) 64 + 28 + 141 = 333
   d) 237 + 542 = 579
   e) 7881 – 3428 = 4453
   f) 9863 – 3619 = 2644

3) a) Use rounding to estimate the answer to these calculations.
   b) Decide if the correct answer will be higher or lower than your estimate.
   c) Then work out the correct answer and check it with your estimate. Were you right?

   The first one has been done for you.

	Estimate	Higher or lower?	Correct answer	Check with estimate
3829 + 2990	6800	The answer will be higher.	6819	Yes, the answer was higher.
3541 – 772				
5403 + 4773				
15 432 – 8762				

⭐ **CHALLENGE!**

Work with a partner. Make up four addition or subtraction problems for each other involving numbers of at least three digits. Use rounding to estimate answers. Work out the answers to your problems and check how close your estimate was to the correct answer.

2 Number – order and place value

2.1 Reading and writing whole numbers

We are learning to read and write five-digit numbers.

Before we start

Miss Wilson called out this number and asked the class to write it in numerals.

nine thousand and ten

Finlay wrote 900010. Explain why Finlay is incorrect. Amman wrote 9100. Explain why Amman is incorrect. Write what Finlay and Amman should have written.

Five-digit numbers are made up of tens of thousands, thousands, hundreds, tens and ones.

Let's learn

We can represent five-digit numbers on place value houses. Each place is worth 10 times more than the place on its right.

Thousands		Ones		
T	O	H	T	O
6	8	2	3	4

We can write five-digit numbers in numerals. We leave a small space between the thousands digit and the other digits. This makes the number easier to read. For example: 68234.

We can write five-digit numbers in words. We use a comma after the word 'thousand' to make the number easier to read. For example: sixty-eight thousand, two hundred and thirty-four.

1) Write the number represented by these place value houses in words.

a)

Thousands		Ones		
T	O	H	T	O
2	4	5	8	6

b)

Thousands		Ones		
T	O	H	T	O
7	2	9	9	8

c)

Thousands		Ones		
T	O	H	T	O
3	1	8	0	0

d)

Thousands		Ones		
T	O	H	T	O
1	0	3	1	2

e)

Thousands		Ones		
T	O	H	T	O
5	5	5	0	5

f)

Thousands		Ones		
T	O	H	T	O
6	0	0	2	0

g)

Thousands		Ones		
T	O	H	T	O
8	9	0	3	0

h)

Thousands		Ones		
T	O	H	T	O
9	0	0	0	0

2) Work with a partner. How many different five-digit numbers can you make by putting two or more of these cards together? Write each number in numerals.

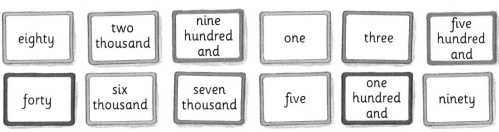

| eighty | two thousand | nine hundred and | one | three | five hundred and |
| forty | six thousand | seven thousand | five | one hundred and | ninety |

1 – 20   Nice try!   21 – 40   Great thinking!   >40   Fantastic!

⭐ CHALLENGE! ....

Nuria thinks that 30 056 says three hundred and fifty-six.
Explain why she is incorrect. Write the number correctly in words.

2 Number – order and place value

2.2 Representing and describing whole numbers

> We are learning to build and describe five-digit numbers.

Before we start

Which models represent the number 2002?

a)

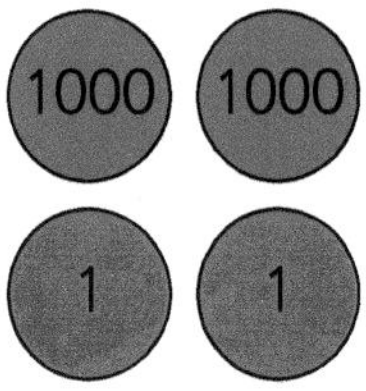

1000 1000

1 1

b)

Thousands		Ones		
	O	H	T	O
	2	2	0	0

c)

| 2 | 0 | 0 | 0 | 0 |

| 2 | 0 |

d)

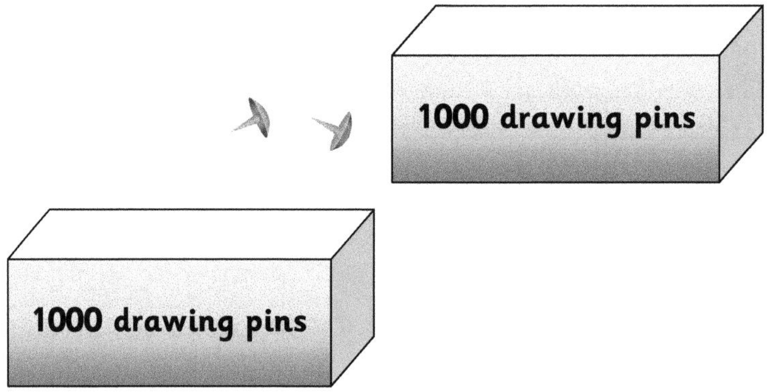

1000 drawing pins

1000 drawing pins

> The position of each digit tells us its place value. The place to the left of a digit is worth 10 times more.

Let's learn

These numbers have the same digits but have different values.

The 2 is worth two hundreds, or 200.

Thousands		Ones		
T	O	H	T	O
6	3	2	7	8

The 2 is worth two lots of ten thousand, or 20 000.

Thousands		Ones		
T	O	H	T	O
2	6	3	8	7

Let's practise

1) Write the value of the digit 5 in each of these numbers. One has been done for you.

a) 45 972 → 5 thousands or 5000

b) 39 285

c) 19 592

d) 79 950

e) 50 024

f) 65 208

2) Nuria made the number 99 099 using these place value arrow cards.

Isla made the number 21 210 using these place value counters.

(10 000) (10 000) (1000) (100) (100) (10)

Represent the numbers below using place value arrow cards or place value counters. Write each number in numerals and draw the cards or counters you used each time.

a) Thirteen thousand, six hundred and sixty-six

b) Forty-four thousand, two hundred and two

c) Seventy thousand and eleven

d) Thirty-six thousand, four hundred

3) Use the digits on these cards to write a five-digit number to match each clue below.

| 5 | 2 | 4 | 1 | 6 |

a) The smallest five-digit number you can make.

b) The largest five-digit number you can make.

c) The largest five-digit number you can make where the 4 is worth 40 000.

d) The smallest five-digit number you can make where the 1 is worth 100.

e) Look carefully at your answers for questions (a) and (b). One of the digits is worth the same in both numbers. Which digit is it and what is it worth?

CHALLENGE!

Finlay is thinking of a five-digit number. Two of the digits are zeros. The hundreds digit is an odd number. The sum of all the digits is 6. What could Finlay's number be? List as many possibilites as you can.

2.3 Place value partitioning of whole numbers

We are learning to partition numbers into tens of thousands, thousands, hundreds, tens and ones.

Before we start

Finlay has made lots of mistakes. Explain why he is wrong in each example then write what he should have written.

a) 3206 = 3000 + 20 + 6

b) 8472 = 8000 + 400 + 20 + 7

c) 2063 = 2000 + 600 + 3

d) 5005 = 500 + 5

Partitioning numbers according to their place values helps us to understand how calculations work.

Let's learn

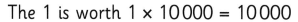

To find the value of a digit, we multiply it by its place.
For example, 13 492

The 1 is worth $1 \times 10\,000 = 10\,000$

The 3 is worth $3 \times 1000 = 3000$

The 4 is worth $4 \times 100 = 400$

The 9 is worth $9 \times 10 = 90$

The 2 is worth $2 \times 1 = 2$

$13\,492 = 10\,000 + 3000 + 400 + 90 + 2$

Thousands		Ones		
T	O	H	T	O
1	3	4	9	2

Let's practise

1) Partition these numbers into ten thousands, thousands, hundreds, tens and ones. One has been done for you.

72 453 = 70 000 + 2 000 + 400 + 50 + 3

a) 14 594   b) 58 234   c) 63 487   d) 27 611

e) 39 925   f) 40 113   g) 78 308   h) 81 075

2) Amman has modelled some five-digit numbers using place value houses. Write down what the digits are worth. One has been done for you.

What is the 1 worth?   10

What is the 3 worth?   30 000

a)

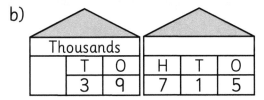

What is the 2 worth?

What is the 8 worth?

b)

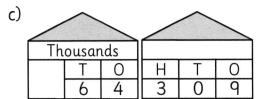

What is the 7 worth?

What is the 3 worth?

c)

What is the 9 worth?

What is the 4 worth?

d)

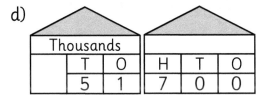

What is the 5 worth?

What is the 1 worth?

3) Partition the numbers shown by each set of base 10 blocks in three different ways. One has been done for you.

2136 can be:

2 thousands, 1 hundred, 3 tens and 6 ones

1 thousand, 11 hundreds, 3 tens and 6 ones (swap 1 thousand for 10 hundreds)

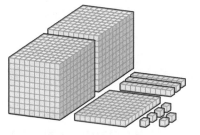

21 hundreds, 3 tens and 6 ones (swap all the thousands for hundreds)

a)

b)

c)

d)

e)

f)

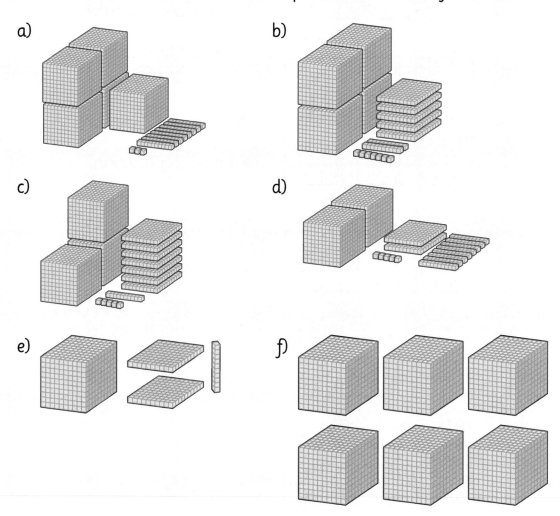

4) Amman is thinking about the number **2135**. He wants to know how many tens and ones it has.

100 tens make 1 thousand so 2 thousand would be 200 tens. There are 10 tens in one hundred so that's 210 tens so far. Add 3 more tens to make 213 tens. So 2135 has 213 tens and 5 ones.

Write these numbers in tens and ones. You may use the diagrams from question 3 to help you.

a) 5083 b) 4527 c) 3714 d) 2285

5) Amman decides to make some four-digit numbers using place value counters but the thousands counters have gone missing! How could he make the following numbers using only 100s, 10s and 1s counters?

a) 1560 b) 1009 c) 6816 d) 9700

CHALLENGE!

Take five cards from a pile of 0–9 digit cards. Use them to make a 'secret number' on your mini-whiteboard. Shuffle the cards and show them to your partner in a different order. Can they guess your secret number? Tell them which place values are correct. Take turns at making and guessing numbers.

Is it 53764?

You have guessed the thousands place and the tens place. Keep guessing!

2 Number – order and place value

2.4 Number sequences

> We are learning to count on and back in 10s, 100s, 1000s and 10 000s.

Before we start

Find the missing numbers. Talk with your teacher or a partner about the patterns you notice.

a)

2690 – ☐ = 2680

2680 – ☐ = 2670

2670 – ☐ = 2660

b)

1499 + ☐ = 1999

1599 + ☐ = 1999

1699 + ☐ = 1999

> The repeating pattern in the way we read and write numbers helps us to understand number sequences.

Let's learn

$10 \times 1 = 10$

10 ones = 1 ten

$10 \times 10 = 100$

10 tens = 1 hundred

$10 \times 100 = 1000$

10 hundreds = 1 thousand

$10 \times 1000 = 10\ 000$

10 thousands = 10 thousand

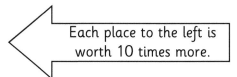

Thousands		Ones		
T	O	H	T	O

×10 ×10 ×10 ×10

Each place to the left is worth 10 times more.

Some number sequences 'bridge' multiples of 10, 100 or 1000. We need to cross the bridge to continue the sequence. For example: 2997, 2998, 2999, **3000**, 3001, 3002, 3003.

1) Write the next five numbers in each sequence. Underline the bridging number.

a) 51 447, 51 448, 51 449

b) 30 894, 30 895, 30 896

c) 76 993, 76 994, 76 995

d) 29 996, 29 997, 29 998

2) Write this number in numerals.
Now write the number that is:

Fifty-nine thousand, eight hundred and ninety

10 more   100 more   1000 more   10 000 more

10 less   100 less   1000 less   10 000 less

3) Copy and complete:

a) 18 990 + 10 =

b) 27 987 + 100 =

c) 89 004 + 1000 =

d) 67 777 + 10 000 =

e) 12 600 – 10 =

f) 91 000 – 100 =

g) 30 000 – 1000 =

h) 15 000 – 10 000 =

CHALLENGE!

Amman's pen has leaked all over his work! Can you identify the missing numbers?

10 800			10 830	10 840		10 860	10 870		10 890
	10 910	10 920	10 930		10 950		10 970	10 980	10 990
				11 040	11 050			11 080	

2 Number – order and place value

2.5 Comparing and ordering whole numbers

We are learning to compare and order five-digit numbers.

Before we start

Nuria is ordering numbers from smallest to largest, but she has made some mistakes. Can you spot which numbers are in the wrong place?

1679, 1769, 679, 6719, 769, 7617, 9671, 967

Write the numbers in the correct order, from smallest to largest, then from largest to smallest.

Look at the first digit of each number and think about its value.

Let's learn

56 349 is less than 95 758 because 50 000 is less than 90 000.

> 56 349 < 95 758 because 50 000 < 90 000

86 248 is greater than 23 002 because 80 000 is greater than 20 000.

> 86 248 > 23 002 because 80 000 > 20 000

The first digit of the number **93 000** is worth 90 000. The first digit of the number **3009** is worth 3000. The first digit of the number **9030** is worth 9000. The first digit of the number **39 003** is worth 30 000.

In order from smallest to largest: 3009, 9030, 39 003, 93 000

Let's practise

1) Copy and complete these statements by writing a symbol in each box. Choose between < and >.

a) 23 290   29 320   b) 10 389   10 144

c) 34 277   73 472   d) 66 137   67 613

e) 58 200   58 020   f) 83 038   83 030

2) a) This table shows the distance by air between Edinburgh and other cities around the world. Write the distances in order from nearest to Edinburgh, to furthest away from Edinburgh.

b) Which is further away from Edinburgh, Rome or Singapore? How much further away is it?

City	Distance from Edinburgh (km)
Amsterdam	661
Auckland	17 905
Barcelona	1666
Delhi	6840
New York	5259
Rome	1930
Singapore	10 930
Sydney	16 880

CHALLENGE!

Make as many five-digit numbers as you can by pairing a yellow card with a white card each time. Write each number in numerals, then compare them using < or >. For example: 49 645 > 49 239.

| Forty-nine thousand | Thirty thousand | Seventy-eight thousand |

| six hundred and forty-five | eight hundred and fifteen | two hundred and thirty-nine |

2 Number – order and place value

2.6 Identifying and placing positive and negative numbers

We are learning to identify and place positive and negative numbers on a number line.

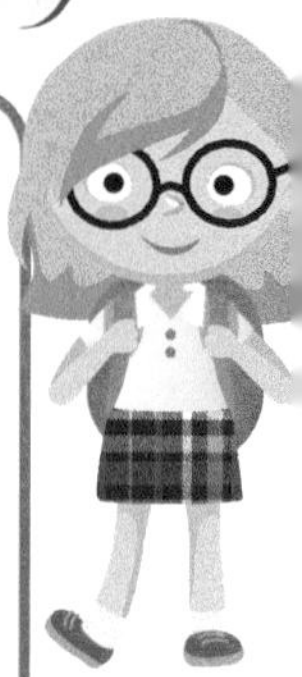

Before we start

	Monday	Tuesday	Wednesday	Thursday	Friday
Day	11°C	7°C	5°C	3°C	1°C
Night	4°C	0°C	–1°C	–5°C	–4°C

Write each temperature in two ways. For example:

Monday

Day: *eleven degrees Celsius* or 11 degrees above freezing

Number lines are useful tools for representing positive and negative numbers.

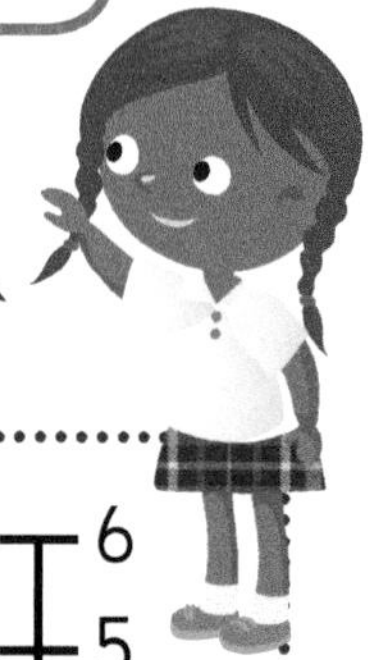

Let's learn

The same number can be displayed on different number lines. On both of these number lines, the blue arrow is pointing to *positive 3* (+3) and the red arrow is pointing to *negative 3* (–3).

–6 –5 –4 –3 –2 –1 0 1 2 3 4 5 6

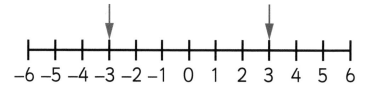

6
5
4
3
2
1
0
–1
–2
–3
–4
–5
–6

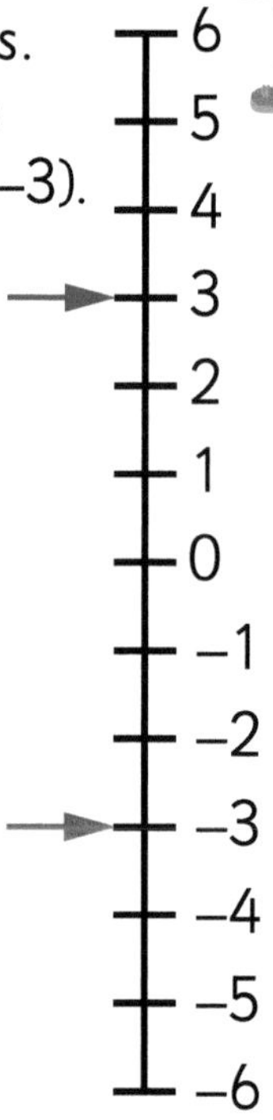

If we subtract a larger number from a smaller number we get a negative number. For example, subtract 5 from 3.

–5 –4 –3 –2 –1 0 1 2 3 4 5

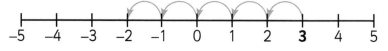

Let's practise

1) Write each label's number in numerals.

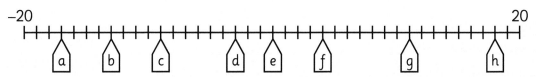

2) Record the temperature shown on each thermometer in both words and numerals. One has been done for you.

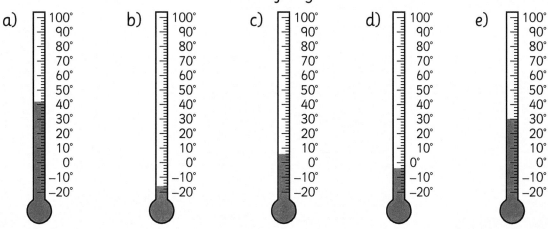

42°C. *Forty-two degrees Celsius.*

3) Write these temperatures in order from coldest to hottest.
   a) −3°C, −8°C, 10°C, −16°C, 0°C, 6°C
   b) 8°C, 14°C, −14°C, −5°C, −7°C, 2°C
   c) −1°C, −49°C, −28°C, −4°C, 0°C, −15°C

CHALLENGE!

The Shanghai Tower is the world's second tallest skyscraper. It has 157 floors above ground level and five floors below ground level. Mr Wang gets into the lift on the 84th floor and travels down 88 floors. Which level is he on now?

Can you find out the name of the world's tallest skyscraper? How many floors does it have? How many floors are below ground level?

Number – order and place value

2.7 Reading and writing decimal fractions

Before we start

Can you spot the 'odd one out' in each row? Justify your thinking.

a) 7 tenths

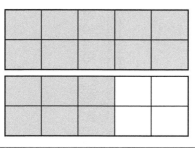

0·7 $\frac{7}{10}$

b) 1 and 6 tenths

1·6 $1\frac{4}{10}$

> We are learning to read and write decimal fractions with hundredths.

> The two digits after a decimal point tell us how many hundredths there are.

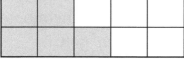

Let's learn

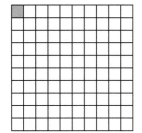

One whole is made up of one hundred **hundredths** $\frac{100}{100} = 1$

One hundredth of this square is shaded.
We write **0·01** and we say *zero point zero one*.

Some decimal fractions have a whole number and a fraction part.

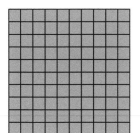
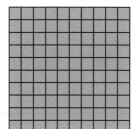
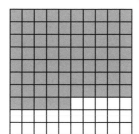

Two whole squares and seventy-five hundredths of a square are shaded.
We write **2·75** and we say *two point seven five*.

Let's practise

1) Write the decimal fraction shown by each diagram in two ways. For example, 0·37 and *zero point three seven*.

a)

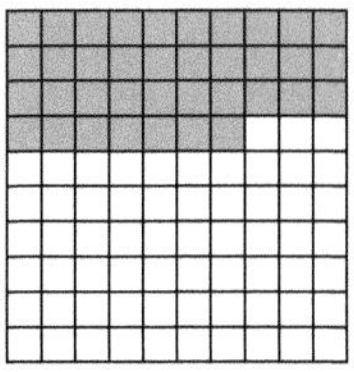

b)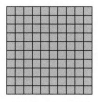

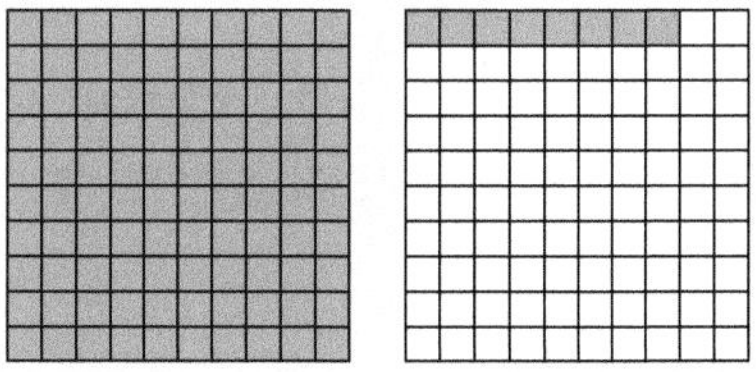

c)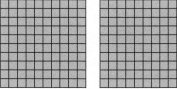

d)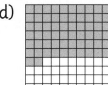

e)

2) This diagram shows four whole squares and sixty-one hundredths of a square, or '*four point six one*'.

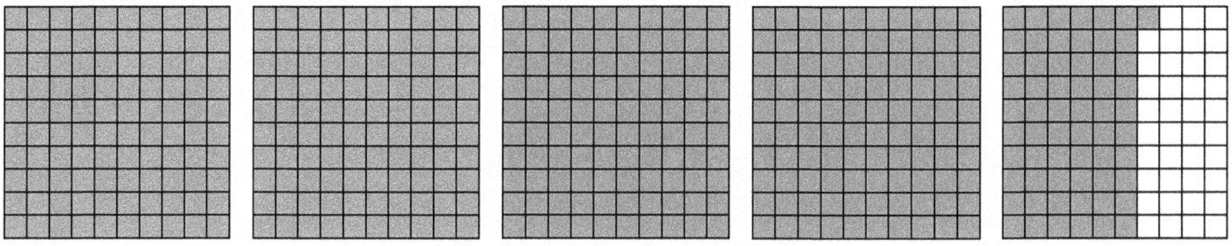

We can write this as the decimal fraction 4·61 or as a **mixed number** $4\frac{61}{100}$

Write in words and as a mixed number.

a) 8·26   b) 7·35   c) 2·02   d) 15·99

 **CHALLENGE!**

Explain why both Nuria and Finlay are incorrect.

8·92

Written as a fraction it says $\frac{82}{9}$.

It says eight point ninety-two.

2.8 Representing and describing decimal fractions

We are learning to build and describe decimal fractions with hundredths.

Before we start

Explain why both models show 2·4

a)

b)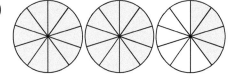

We can represent decimal fractions in different ways.

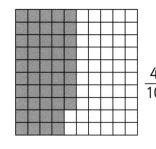

Let's learn

The fraction part of a decimal fraction represents part of a whole.

Numbers smaller than 1 can be written as both a fraction and as a decimal fraction. For example:

$\frac{48}{100}$ or 0·48

Let's practise

1) a) Explain why each model represents the decimal fraction 1·05.

i)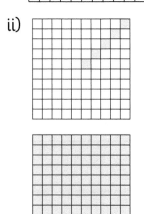

ii)   iii)

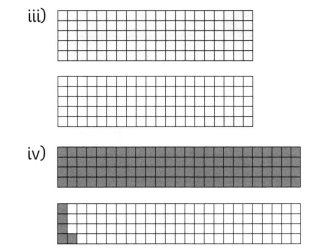

iv)

b) What fraction of each model is **unshaded**? Write your answer as a fraction and as a decimal fraction.

2) Represent each of these decimal fractions on squared paper.

    a) 2·07   b) 0·56   c) 1·19   d) 0·04   e) 6·83   f) 3·50

3) Amman uses these place value arrow cards to represent the decimal fraction 4·68

Make these decimal fractions using place value arrow cards. Draw the place value arrow cards you use each time.

    a) 5·11   b) 7·09   c) 0·22   d) 9·65
    e) 2·41   f) 8·03   g) 1·70   h) 3·33

CHALLENGE!

1) How would you use a grid divided in 100 parts to show eight tenths?

2) How would you use a grid divided in 100 parts to show 0·6?

2.9 Comparing and ordering decimal fractions

We are learning to place decimal fractions with two decimal places on a number line.

Before we start

Some numbers from the decimal tenth number line have been stolen! Write down as many ways as possible that Amman could describe each missing number.

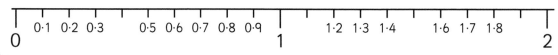

We can compare and order decimal fractions by thinking about the value of each digit.

Let's learn

To compare decimal fractions with two decimal places we begin by looking at the whole number. If two decimal fractions share the same whole number we only need to compare the fractional parts to say which is larger and which is smaller. For example:

2·04 < 2·24 because 4 hundredths is less than 24 hundredths.

10·91 > 10·19 because 91 hundredths is more than 19 hundredths.

A number line can help us think about the order of the digits. Work with a partner to identify the missing decimal fractions on this number line.

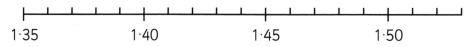

Let's practise

1) Write the decimal fraction that each arrow is pointing to.

a)

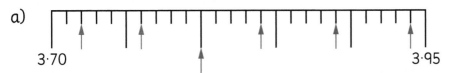

3·70   3·95

b)

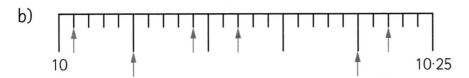

10   10·25

2) True or false?
   a) 20·62 > 20·26   b) 15·51 = 51·15   c) 9 < 7·24
   d) 18 > 18·00   e) 3·31 < 3·13   f) 38·11 > 38
   g) 106·45 < 106·73   h) 299·43 > 299·03

3) Look again at the false statements in question 2. Change the symbol in each example to make the statement true.

4) Copy these number sequences. Replace each box with a whole number to keep the order correct.

   a) ☐, 2·69, 3·21, 3·84, ☐, 8·33, 8·35, ☐, 9·01, 9·41

   b) 13·33, 11·31, ☐, 10·77, ☐, ☐, 2·14, 1·42, ☐

CHALLENGE!

Write these numbers in order from smallest to largest:

9   19   9·21   1·92   1·29   9·99   1·9

Now make your own set of decimal number cards. Challenge a friend to arrange them in the correct order from smallest to largest, or from largest to smallest.

3 Number – addition and subtraction

3.1 Mental addition and subtraction

We are learning to add and subtract whole numbers mentally.

Before we start

Calculate mentally. Explain how you worked each answer out.

a) 51 + 333   b) 514 + 46   c) 22 + 268

d) 696 – 79   e) 363 – 28   f) 197 – 49

If you transform something, you change it.

Let's learn

We can transform a calculation by changing the numbers, so that they are easier to work with, without changing the value of the answer. For example **137 + 125**:

140 + 122

240

260

262

I changed the calculation to 140 + 122

I added 3 to 137 and took 3 away from 125.

I jotted down some of the numbers to help me keep track of my thinking.

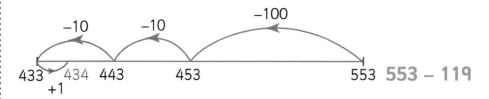

553 – 119

Instead of 553 – 119 I did 554 – 120.
554 take away 100 is 454, take away 20 is 434.
I checked that my strategy worked by working
out 553 – 119 on an empty number line.

1) Transform the numbers in these additions then calculate each answer mentally. You may find jottings helpful.

a) 326 + 159
b) 274 + 288
c) 437 + 369
d) 183 + 599
e) 306 + 258
f) 793 + 419
g) 564 + 428
h) 837 + 859
i) 613 + 908
j) 996 + 219

Compare answers with a friend and agree the correct solutions. Match each correct answer with a letter in the table to reveal a secret word.

806	562	992	1215	1212	485	1521	564	1696	782
R	T	G	S	E	S	E	E	T	A

2) Transform the numbers in these subtractions then calculate each answer mentally. You may find jottings helpful.

a) 862 – 429
b) 521 – 116
c) 987 – 767
d) 1838 – 211
e) 2745 – 599
f) 4567 – 122
g) 1148 – 298
h) 7853 – 5536
i) 3797 – 2000

Check your answers using an empty number line.

3) True or false? Explain your thinking.

a) 3852 + 2374 = 3853 + 2375
b) 1789 + 6552 = 6552 + 1789
c) 8999 – 4999 = 9000 – 5000
d) 4501 – 1000 = 4105 – 1000

CHALLENGE!

Explain why Nuria is incorrect. Change the tens and ones digits of the number 1252 to make the statement true. Write down the true statement.

I think 2741 – 1263 = 2742 – 1252

3.2 Adding and subtracting a string of numbers

We are learning to add and subtract four-digit numbers mentally by making multiples of 10, 100 or 1000.

Before we start

Finlay is calculating the answers to the questions in bold but he has made some mistakes:

228 + 365 + 42
228 add 42 is 270
365 add 270
is 5135

825 – 174 – 526
526 add 174 is 690
825 subtract 690
is 135

Fix Finlay's errors to find the correct answer to each calculation.

Let's learn

Looking for multiples of 10, 100 or 1000 can make mental addition and subtraction easier.

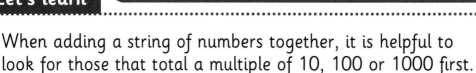

When adding a string of numbers together, it is helpful to look for those that total a multiple of 10, 100 or 1000 first. For example:

513 + 1047 + 200

47 +   13 =   60
560 + 1000 = 1560
1560 +   200 = 1760

4340 + 260 + 2216

340 +   260 =   600
4600 + 2216 = 6816

5310 + 1690 + 809

310 +   690 = 1000
5000 + 1000 + 1000 = 7000
7000 +   809 = 7809

Let's practise

1) Calculate mentally. Look for multiples of 10, 100 or 1000. Make jottings to help you keep track of your thinking.

   a) 329 + 211 + 1328   b) 101 + 1020 + 780
   c) 1165 + 535 + 3151   d) 3000 + 1590 + 410
   e) 5000 + 2099 + 2000   f) 2000 + 4800 + 1200
   g) 5000 + 1269 + 3000   h) 1860 + 3177 + 1003
   i) 4260 + 1288 + 3040   j) 3320 + 1272 + 5080

> When subtracting a string of numbers, it is easier to look for those that total a multiple of 10, 100 or 1000 and subtract them together rather than separately. For example:
>
> | 9900 – 452 –148 | 452 plus   148 equals   600 |
> |---|---|
> | | 9900 minus 600 equals   9300 |

2) Calculate mentally. Look for multiples of 10, 100 or 1000. Make jottings to help you keep track of your thinking.

   a) 5005 – 200 – 800   b) 7818 – 470 – 130
   c) 6921 – 350 – 550   d) 9760 – 1100 – 2900
   e) 4860 – 1234 – 1426   f) 8888 – 2198 – 2202
   g) 7000 – 3760 – 1240   h) 9001 – 1800 – 5200
   i) 10 000 – 2670 – 1130   j) 13 100 – 3000 - 7000

3) Find the missing number in each calculation.

   a) 3000 + ☐ + 960 = 6700   b) 4598 – 1000 – ☐ = 3321

CHALLENGE!

Fill in the missing digits. Write the completed number sentences.

1✱✱6 + 3200 + 1240 = 5446   4265 – 1065 – ✺ ✺ ✺ = 3000

3.3 Adding and subtracting multiples of 10, 100 and 1000

We are learning to use place value to add and subtract multiples of 10, 100 and 1000 to and from five-digit numbers.

Before we start

Fill in the missing digits to make each number sentence true:

a) $8214 + 1\square\square\square = 9516$

b) $2730 + \square000 = 9\square\square\square$

c) $5049 - \square\square\square\square = 1003$

d) $\square\square\square\square - 3121 = 4656$

Place value can help us to add and subtract multiples of 1000 to and from a five-digit number.

Let's learn

Nuria partitions each number into its place values to work out **24866 + 1900**.

First she splits 24866 into **24000 + 800 + 60 + 6**. Then she splits 1900 into **1000 + 900**.

She adds **24000 + 1000 + 800 + 900 + 60 + 6**

$\qquad$ = **25000 + 1700 + 60 + 6**

$\qquad$ = **26000 + 700 + 60 + 6**

$\qquad$ = **26766**

Finlay uses number facts and place value to work out **35 879 – 14 420**.

35 thousands take away 14 thousands equals 21 thousands.

8 hundreds take away 4 hundreds equals 4 hundreds.

7 tens take away 2 tens equals 5 tens.

9 take away 0 equals 9.

So 35 879 − 14 420 = 21 459.

Thousands		Ones		
T	O	H	T	O
2	1	4	5	9

Let's practise

1) Use Nuria's strategy to work out:
   a) 28 332 + 3590
   b) 49 870 + 1710
   c) 62 113 + 9990
   d) 30 986 + 10 560
   e) 61 789 + 13 650
   f) 96 804 + 10 430
   g) 24 680 − 19 540
   h) 76 298 − 12 100
   i) 99 555 − 81 300

2) Use Finlay's strategy to work out:
   a) 14 670 − 9510
   b) 39 487 − 3270
   c) 67 721 − 9100
   d) 25 599 − 10 000
   e) 89 936 − 29 620
   f) 41 814 − 18 701
   g) 60 561 + 29 430
   h) 77 100 + 18 660
   i) 86 050 + 57 845

3) Replace ♦ with a multiple of 100 or 1000 to make each number sentence true.
   a) 79 119 = ♦ + 76 119
   b) 35 898 = 34 298 + ♦
   c) 13 600 + 19 300 + 34 000 = ♦
   d) 44 700 = ♦ − 11 000
   e) 26 090 = 30 090 − ♦
   f) 99 000 − 22 000 − 15 000 = ♦

CHALLENGE!

Amman thinks the answer to Mrs Ferguson's question is 44 500.

What is 54 200 minus 10 700?

Explain why Amman is incorrect. Work out the correct answer to Mrs Ferguson's question. You may find it helpful to draw an empty number line.

3 Number – addition and subtraction

3.4 Using place value partitioning

We are learning to add and subtract four- and five-digit numbers by partitioning them into thousands, hundreds, tens and ones.

Before we start

Fill in the missing numbers then calculate the answers.

a)
```
    848
+   484
  1 2 0 0
  ✳✳✳
      1 2
```

b)
```
    766
+   395
  ✳✳✳✳
    1 5 0
      1 1
```

c)
```
    279
+   558
    7 0 0
  ✳✳✳
    ✳✳
```

Partitioning helps us to calculate with large numbers by focusing on the place values one at a time.

Let's learn

Finlay is working out the answers to 33 478 + 45 376 and 57 854 – 31 634 by partitioning.

```
                        3 3 4 7 8
                    +   4 5 3 7 6
Add the tens of thousands  7 0 0 0 0
Add the thousands          8 0 0 0
Add the hundreds             7 0 0
Add the tens                 1 4 0
Add the ones                  1 4
                        7 8 8 5 4
```

Writing the numbers down in columns helps me to see the partitions more clearly.

$$
\begin{array}{r}
5\ 7\ 8\ 5\ 4 \\
-\ 3\ 1\ 6\ 3\ 4 \\
\hline
\end{array}
$$

Subtract the tens of thousands　2 0 0 0 0
Subtract the thousands　6 0 0 0
Subtract the hundreds　2 0 0
Subtract the tens　2 0
Subtract the ones　0

$$\overline{2\ 6\ 2\ 2\ 0}$$

I need to line each digit up according to its place value.

Let's practise

1) Use Finlay's method to calculate the answers to these additions.

　　a) 4728 + 5258　　　b) 5757 + 6309　　　c) 9653 + 4685
　　d) 19 875 + 6421　　e) 35 251 + 5674　　f) 22 223 + 3845
　　g) 62 418 + 12 426　h) 51 963 + 25 927　i) 46 345 + 46 861

2) Use Finlay's method to calculate the answers to these subtractions.

　　a) 28 926 – 4701　　b) 17 569 – 1332　　c) 35 967 – 2615
　　d) 67 324 – 16 210　e) 29 798 – 12 526　f) 85 647 – 71 336
　　g) 56 986 – 36 271　h) 90 766 – 80 510　i) 72 873 – 62 661

3) A palindrome is a number that reads the same forwards and backwards, for example 3443. Complete each number sentence so that each total is a palindrome.

　　a) ✳000 + 100 + 10 + 7　　b) 8000 + ✳00 + 590 + 8
　　c) 5000 + 200 + 1✳0 + 5　　d) 2000 + 3✳0 + 340 + ✳

CHALLENGE!

Using each digit only once each time write two 4-digit numbers that when added give:

a) the largest total　　　b) the second largest total

| 4 | 8 | 2 | 7 | 1 | 0 | 3 | 5 |

3 Number – addition and subtraction

3.5 Adding four-digit numbers using standard algorithms

We are learning to add four-digit numbers using a standard written method.

Before we start

Use two or more cards to make the number 1975 in five different ways. Cards may be used more than once. Not all cards may be needed.

| 1 thousand | 19 tens | 7 tens | 19 hundreds |

| 9 hundreds | 97 tens | 19 tens | 975 ones |

| 957 ones | 75 ones | 5 ones |

Algorithms can help us with addition calculations that are too tricky to work out mentally.

Let's learn

An algorithm is a set of instructions for carrying out a calculation. The steps must be carried out in the correct order.

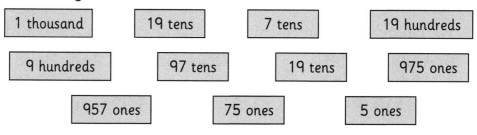

```
  1 1 1
  2 7 5 9
+ 3 6 8 6
  6 4 4 5
```

9 ones + 6 ones = 15 ones = 1 ten and 5 ones. Write 5 in the ones column and **carry** 1 ten over into the tens column.

1 ten + 5 tens + 8 tens = 14 tens = 1 hundred and 4 tens. Write 4 in the tens column and **carry** 1 hundred.

There are 14 hundreds which is the same as 1 thousand and 4 hundreds. Write 4 in the hundreds column and **carry** 1 thousand into the thousands column.

Add the thousands. 1 thousand + 2 thousands + 3 thousands = 6 thousands.

Let's practise

1) Write each calculation as a column addition. Use Isla's method to find the answers.

    a) 7255 + 1938   b) 4872 + 2949   c) 3684 + 4056
    d) 2756 + 3765   e) 1466 + 9295   f) 5756 + 4366
    g) 8647 + 6478   h) 6285 + 6591   i) 9507 + 4856

2) Calculate the answers using the standard written algorithm for addition.

    a) 3846 + 1638 + 2647   b) 2752 + 2338 + 8464
    c) 5868 + 6415 + 2968   d) 5527 + 5255 + 5635
    e) 7666 + 2378 + 4444   f) 3960 + 1076 + 5924

3) Now try these additions. Think carefully about the place values of the digits.

    a) 2468 + 947 + 1875   b) 635 + 8352 + 2158
    c) 3816 + 9767 + 406   d) 38 + 485 + 1673 + 3487

4) Find the missing number on each brick by adding together the two numbers directly below it.

1462	2319	1458	2629	1229

⭐ **CHALLENGE!**

Fill in the missing digits.

a)
```
    4 7 8 6
  + 2 ✳ 3 ✳
  ✳ 7 ✳ 7
```

b)
```
    3 7 5 5
  + ✳ ✳ ✳ ✳
    6 4 4 4
```

c)
```
    ✳ ✳ ✳ ✳
  +   4 1 7 4
    1 1 6 6 2
```

d)
```
    5 ✳ 2 ✳
  + ✳ 2 9 3
    1 3 1 ✳ 1
```

3 Number – addition and subtraction

3.6 Column subtraction: three-digit numbers

We are learning to subtract three-digit numbers using a semi-formal written method.

Before we start

Amman is counting back in ones but has made some mistakes. Write what Amman should have written.

a)
If I start on 8 and take 10 jumps backwards I will be on 2.

b)
If I start on 3 and take 7 jumps backwards I will be on –3.

Isla is counting back in tens but has also made some mistakes. Write what Isla should have written.

c)
If I start at 50 and count back 80 I will be on –20.

d)
If I start at 10 and count back 70 I will be on 60.

Sometimes it is easier to subtract one number from another if we set the numbers down in a column.

Let's learn

Writing numbers down in columns can help us with subtractions that are too tricky to do mentally. This is because columns make it easier to look at the ones, the tens and the hundreds separately.

```
      9 2 8
    – 2 6 5
```
Subtract the hundreds   7 0 0
Subtract the tens   – 4 0
Subtract the ones   3
```
      6 6 3
```

900 take away 200 is 700
20 take away 60 is minus 40
8 take away 5 is 3
703 take away 40 is 663

```
      8 7 1
    – 4 6 7
```
Subtract the hundreds   4 0 0
Subtract the tens   1 0
Subtract the ones   – 6
```
      4 0 4
```

800 take away 400 is 400
70 take away 60 is 10
1 take away 7 is minus 6
410 take away 6 is 404

Let's practise

1) Use Nuria's method to calculate:
   a) 883 – 374   b) 956 – 528   c) 658 – 477   d) 307 – 173
   e) 746 – 165   f) 842 – 615   g) 562 – 253   h) 438 – 264

2) Now find the answers to these subtractions using the column method.
   a) 683 – 389   b) 744 – 487   c) 513 – 439   d) 825 – 176
   e) 414 – 338   f) 978 – 589   g) 636 – 287   h) 818 – 379
   i) 676 – 481   j) 753 – 338   k) 502 – 416   l) 903 – 518

CHALLENGE!

Finlay has tried to work out the answer to 635 – 254 but has made a mistake. Can you see where he has gone wrong? Write what Finlay should have written.

```
      6 3 5
    – 2 5 4
      4 0 0
        2 0
          1
      4 2 1
```

3.7 Column subtraction: four-digit numbers

> We are learning to subtract four-digit numbers using a semi-formal written method.

Before we start

Count back in hundreds on the number line to find:

a) 200 take away 300

b) 400 take away 900

+———+———+———+———+———+———+———+———+———+———+
−500 −400 −300 −200 −100   0   100   200   300   400   500

Now try these:

c) 200 take away 800

d) 100 take away 800

e) 200 take away 900

f) 600 − 900

> We can use the same semi-formal written method that we used to subtract three-digit numbers for subtractions involving four-digit numbers.

Let's learn

```
        7 5 5 6
      − 2 8 3 1
```
Subtract the thousands 5 0 0 0
Subtract the hundreds   − 3 0 0
Subtract the tens   2 0
Subtract the ones   5
                       ————
                       4 7 2 5

> 7000 take away 2000 is 5000
> 500 take away 800 is minus 300
> 50 take away 30 is 20
> 6 take away 1 is 5
> That's 5025 take away 300
> The answer is 4725

$$\begin{array}{r} 8823 \\ -\ 4576 \\ \hline \end{array}$$

Subtract the thousands   4 0 0 0
Subtract the hundreds   3 0 0
Subtract the tens   – 5 0
Subtract the ones   – 3
                             4 2 4 7

8000 take away 4000 is 4000
800 take away 500 is 300
20 take away 70 is minus 50
3 take away 6 is minus 3
That's 4300 take away 53
The answer is 4247

Let's practise

1) Use the boys' method to calculate:

     a) 5562 – 2347   b) 7264 – 2854   c) 8626 – 5362
     d) 8461 – 4467   e) 4736 – 1262   f) 9185 – 1664
     g) 6278 – 3854   h) 3927 – 1565   i) 2571 – 1426

2) Find the answers to these subtractions using the column method.

     a) 7114 – 2563   b) 4785 – 1967   c) 3813 – 1675
     d) 9425 – 3663   e) 5785 – 3866   f) 9583 – 4648
     g) 8218 – 6877   h) 6258 – 3481   i) 3251 – 1554
     j) 7311 – 3334   k) 4256 – 2858   l) 6243 – 5657

CHALLENGE!

Now use the boys' strategy to work out the answers to these subtractions. What do you notice?

     a) 8362 – 5483   b) 9747 – 8888   c) 7433 – 1685
     d) 4321 – 2565   e) 5231 – 2879   f) 3026 – 1468

3.8 Subtracting three-digit numbers using standard algorithms

We are learning to subtract three-digit numbers using a standard written method.

Before we start

Pencils are sold in boxes of 100, packs of 10 or singly.

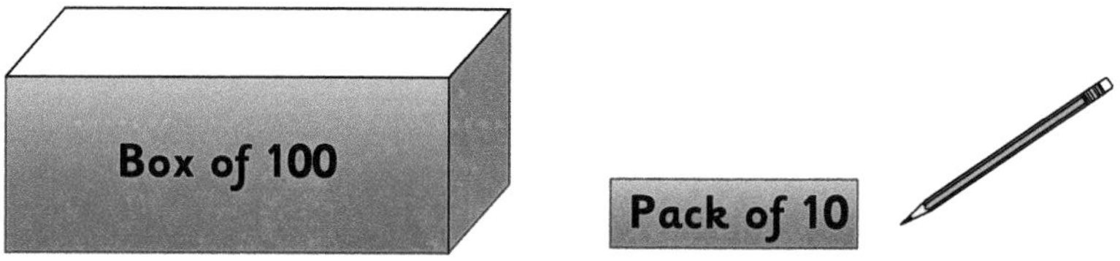

Box of 100

Pack of 10

Fill in the missing numbers to show the different ways that Miss Mackay can buy 234 pencils for the school.

☐ boxes of 100, three packs of 10 and ☐ single pencils

one box of 100, ☐ packs of 10 and ☐ single pencils

two boxes of 100, two packs of 10 and ☐ single pencils

Algorithms can help us with subtraction calculations that are too tricky to work out mentally.

Isla, Finlay and Amman are using the subtraction algorithm.
Discuss each example with a partner or as a group.

```
  4 1
4 5̷ 1̷
- 1 3 7
―――――
  3 1 4
```

I can't take 7 away from 1.
Exchanging 1 ten for 10 ones will give me
4 tens and 11 ones.
11 take away 7 is 4.
4 tens take away 3 tens leaves 1 ten.
4 hundreds − 1 hundred = 3 hundreds.

```
  6 1
7̷ 3 8
- 4 8 6
―――――
  2 5 2
```

8 take away 6 is 2.
I can't take 8 tens from 3 tens.
Exchanging 1 hundred for 10 tens will give me
6 hundreds and 13 tens.
13 tens take away 8 tens leaves 5 tens.
6 hundreds − 4 hundreds = 2 hundreds.

```
  4 11 1
5̷ 2̷ 4̷
- 2 7 8
―――――
  2 4 6
```

I can't take 8 away from 4.
Exchanging 1 ten for 10 ones will
give me 1 ten and 14 ones.
14 take away 8 is 6.
I can't take 7 tens away from 1 ten.
Exchanging 1 hundred for 10 tens will give
me 4 hundreds and 11 tens.
11 tens take away 7 tens leaves 4 tens.
4 hundreds − 2 hundreds = 2 hundreds.

Let's practise

1) Use Isla or Finlay's method to find:

    a) 781 – 239   b) 635 – 293   c) 808 – 585   d) 724 – 632
    e) 562 – 328   f) 407 – 186   g) 326 – 184   h) 666 – 272

2) Use Amman's method to find:

    a) 632 – 354   b) 872 – 483   c) 527 – 388   d) 935 – 467
    e) 422 – 176   f) 916 – 568   g) 744 – 267   h) 635 – 279

3) Now calculate the answers to these questions using the subtraction algorithm. Think carefully. Will you need to **exchange** once or twice?

    a) 314 – 135   b) 442 – 286   c) 982 – 219   d) 553 – 496
    e) 821 – 624   f) 936 – 369   g) 712 – 387   h) 637 – 570
    i) 432–378   j) 536 – 328   k) 632 – 386   l) 917 – 272

4) Match each answer in question 3 to a capital letter from the table below to reveal a secret word.

567	645	156	57	208	246	67	179	54	763	325	197
L	S	A	C	O	N	T	C	I	L	A	U

CHALLENGE!

What could the missing digits be?
Can you find all of the possibilities?

$$
\begin{array}{r}
4\ 3\ 6 \\
-\ \text{❋}\ \text{❋}\ 9 \\
\hline
1\ \text{❋}\ 7 \\
\end{array}
$$

3 Number – addition and subtraction

3.9 Subtracting four-digit numbers using standard algorithms

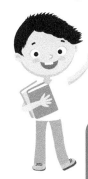

We are learning to subtract four-digit numbers using a standard written method.

Before we start

Find four different ways to make 7360 using two or more of these cards. Cards may be used more than once.

- 73 tens
- 6 thousands
- 3 hundreds
- 36 tens
- 60 ones
- 6 tens
- 13 hundreds
- 36 ones
- 7 thousands

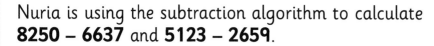

Algorithms can help us with subtraction calculations that are too tricky to work out mentally.

Let's learn

Nuria is using the subtraction algorithm to calculate **8250 – 6637** and **5123 – 2659**.

```
   7 1 4                        4 10 11
   8 2 5 0                      5 1 2 3
 –   6 6 3 7                  –   2 6 5 9
   ─────────                    ─────────
   1 6 1 3                      2 4 6 4
```

I can't take seven away from zero. If I **exchange one ten for 10 ones** I will have 4 tens **and** 10 ones.
10 ones take away 7 ones leaves 3 ones.
4 tens take away 3 tens leaves 1 ten.
I can't take 6 hundreds away from 2 hundreds.
If I **exchange one thousand for 10 hundreds** I will have
7 thousands **and** 12 hundreds.
12 hundreds take away 6 hundreds leaves 6 hundreds.
7 thousands take away 6 thousands leaves 1 thousand.
The answer is 1613.

I can't take 3 away from 9. If I **exchange one ten for 10 ones** I will have 1 ten **and** 13 ones. 13 take away 9 is 4.
I can't take 5 tens away from 1 ten. If I **exchange one hundred for 10 tens** I will have 0 hundreds **and** 11 tens.
11 tens take away 5 tens leaves 6 tens.
I can't take 6 hundreds away from 0 hundreds. If I **exchange one thousand for 10 hundreds** I will have
4 thousands **and** 10 hundreds.
10 hundreds take away 6 hundreds leaves 4 hundreds.
4 thousands take away 2 thousands leaves 2 thousands.
The answer is 2464.

Let's practise

1) Use Nuria's method to find the answers to these subtractions.

   a) 5522 – 1427   b) 7166 – 6248   c) 3921 – 1960
   d) 5360 – 4515   e) 7040 – 516   f) 3964 – 1372
   g) 8291 – 1631   h) 8127 – 676   i) 9367 – 1538

Think carefully about where you need to **exchange**.

2) Calculate the answers to these questions using the subtraction algorithm.

   a) 8123 – 4727   b) 7135 – 1217   c) 6441 – 5766
   d) 8932 – 5469   e) 6102 – 1977   f) 5322 – 1436
   g) 2196 – 529   h) 1514 – 736   i) 3175 – 1688

3) Now try these.

   a) 4206 – 1147   b) 5031 – 2456
   c) 7040 – 3178   d) 9204 – 6828

More Challenging!

CHALLENGE!

The same four-digit number is needed to make both these algorithms correct. Find the missing number.

```
  8 2 4 1              6 8 4 8
– ✹ ✹ ✹ ✹           – ✹ ✹ ✹ ✹
 ─────────            ─────────
  2 4 7 9              1 0 8 6
```

3.10 Mental and written strategies

We are learning to choose an appropriate strategy to solve an addition or subtraction problem.

Before we start

Work out the answers to these calculations using a strategy of your choice. What strategy did you use and why?

a) 5040 + 1206   b) 3009 – 2992

There are many strategies that we can use to solve number problems.

Let's learn

A good strategy will help you to solve a problem efficiently. Whose strategy do you think is the most efficient way to find **1459 + 328**?

I used the addition algorithm.

$$
\begin{array}{r}
1\overset{1}{4}59 \\
+\ 328 \\
\hline
1787
\end{array}
$$

I used round and adjust. I did 1460 + 327 which is 1000 + 700 + 80 + 7.

I drew an empty number line.

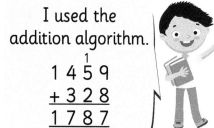

+300   +20   +8

1459   1759   1779   1787

I partitioned. I wrote the numbers down in a column.

$$
\begin{array}{r}
1459 \\
+\ 328 \\
\hline
1000 \\
700 \\
70 \\
17 \\
\hline
1787
\end{array}
$$

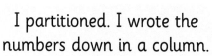
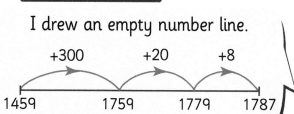
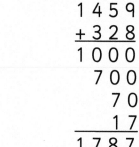

Let's practise

1) Find the missing number on each brick of these addition pyramids by adding together the two numbers directly below it. Think carefully about your choice of strategy.

352	639	578	449	659

2) Work with a partner. Take turns to roll four dice to create a four-digit number and subtract the number from 9999. Continue taking turns to roll numbers and subtract from your previous answer. The first player to reach an even number under 1000 is the winner. Think carefully about your choice of strategy.

Player One	Player Two
9999 – [ ] = [ ]	9999 – [ ] = [ ]

CHALLENGE!

4830	3455	1435	3445	2840	4820

a) Find the difference between the largest and the smallest number.

b) Which two numbers have a difference of 1365?

c) Which three numbers total 9105?

3 Number – addition and subtraction

3.11 Representing and solving word problems

We are learning to represent and solve the same word problem in different ways.

Before we start

Represent this word problem as a bar model, on an empty number line and as an algorithm.

The Glasgow Science Centre received 4785 visitors during the first week of July and 5162 visitors during the second week of July. How many people visited over the two weeks?

Representing the same problem in different ways helps to deepen our understanding.

Let's learn

Nuria and Amman draw a Think Board to help them think about this word problem.

The first ever adding machine was invented by Blaise Pascal and Wilhelm Schickard in 1642. How many years ago was this?

Word problem	Bar model
The first ever adding machine was invented by Blaise Pascal and Wilhelm Schickard in 1642. How many years ago was this?	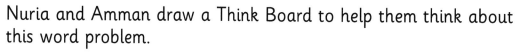

Empty number line	Calculation
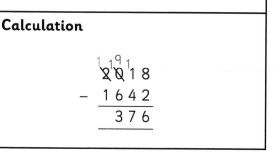	

1) Create Think Boards for these word problems. Represent each problem as a bar model, as an empty number line and as an algorithm.

First telephone invented in 1876

First aeroplane invented in 1903

First petrol powered car invented in 1887

Adding machine invented in 1642

First television invented in 1924

First lightbulb invented in 1879

a) How many years after the telephone was the television invented?

b) The first thermometer was invented in 1593. How many years before Pascal and Schickard's adding machine is this?

c) The steam piston engine was invented in 1712, one hundred and sixty years before the petrol engine. When was the petrol engine invented?

2) Represent each word problem as a bar model, then calculate each answer mentally.

a) How many years after the petrol engine was the first petrol powered car invented? (You need one answer from question 1).

b) Which was invented first, the aeroplane or the petrol-powered car? How many years were there between these two inventions?

c) How many years ago was the light bulb invented?

CHALLENGE!

The first adding machine was invented in 1642. The first vacuum cleaner was invented in 1901. Isla thinks the vacuum cleaner was invented 251 years after the adding machine. Explain why Isla is incorrect. Work out the correct answer to this problem.

3.12 Solving multi-step word problems

We are learning to choose the most appropriate strategy to solve a multi-step word problem.

Before we start

Solve this word problem using a strategy of your choice.

Mrs Henry buys 2200 pencils for her school.
The school uses 306 pencils during the first term,
417 during the second term and 629 during the third
term. How many pencils does Mrs Henry have left?

The same word problem can be solved in more than one way.

Let's learn

Amman and Finlay draw a bar model to help them think about this problem.

A farmer puts his herd of 1036 cows into four fields. He puts 258 in North field, 378 in South field and divides the rest equally between East and West fields. How many cows are in each of East and West fields?

1036			
258	378	?	?

I partitioned
$500 + 120 + 16 = 636$.

I wrote an algorithm for 258 + 378 and got the answer 636.

So there must be 400 cows in East and West fields.

The number of cows in East and West fields is the same. 400 divided by 2 equals 200.

Let's practise

1) There are 12 000 tickets for the Rugby Cup Final. Sport Event sells 7100 between Monday and Friday and 1450 on **each of** Saturday and Sunday. How many tickets are left unsold?

2) Finlay's family is planning a holiday costing £3217. They have saved £1855 so far. They know that they will get £250 off for booking online. How much money do they still need to save?

3) Amman's brother has saved £4020. He buys a car costing £3476, spends £120 on road tax and £367 on insurance. Does he have enough money left to buy two seat covers costing £29 each?

4) Nuria's family is moving house and decide to buy some new furniture. They buy three TVs, each costing £279, for Nuria and her brothers' rooms. They also buy a new sofa costing £1195, a bed costing £374 and a washing machine costing £459. How much did they spend?

5) 10 000 concert tickets went on sale on Monday. By Thursday, all 10 000 tickets had been sold. 3563 were bought on Monday and 2897 were bought on Tuesday. 1000 more tickets were bought on Wednesday than on Thursday. How many tickets were bought on Wednesday?

CHALLENGE!

Write your own word problem. Ask a partner to draw a Think Board to represent and solve it. Your partner should have to do at least two calculations to find the solution.

3.13 Adding decimal fractions and whole numbers

We are learning to add whole numbers and decimal fractions with one decimal place.

Before we start

Copy and complete each number line.

a)

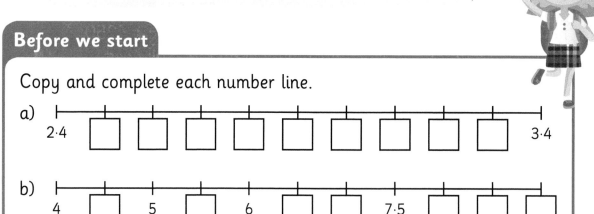

b)

Understanding the value of each digit helps us to add whole numbers and decimal fractions.

Let's learn

Finlay uses **'round and adjust'** to find the answer to **4·5 + 199**.

He knows that 4·5 means 4 ones and 5 tenths. He rounds 199 up to 200 and **compensates** by taking one away from 4·5. His calculation now says **3·5 + 200**.

> 4·5 + 199 is the same as 3·5 + 200. The answer is 203·5.

Isla is working out **13·9 + 5·3**.

She **rounds** 13·9 up to the nearest whole number by adding one tenth **then compensates** for this by subtracting one tenth from 5·3. Her calculation now says **14 + 5·2**.

> 13·9 + 5·3 is the same as 14 + 5·2. The answer is 19·2.

Nuria is working out the answer to **40·2 + 3·7**.

She **partitions** each decimal fraction into whole numbers and tenths.

40·2 + 3·7 = 40 + 3 + 0·2 + 0·7 = 43 + 0·9 = 43·9

Let's practise

1) Use Finlay's strategy to calculate:
   a) 59 + 8·4
   b) 7·3 + 39
   c) 98 + 6·1
   d) 11·6 + 499
   e) 169 + 13·7
   f) 12·9 + 129
   g) 68 + 19·5
   h) 23·7 + 48

2) Use Isla's strategy to calculate:
   a) 7·2 + 11·9
   b) 62·9 + 1·7
   c) 45·3 + 1·9
   d) 12·8 + 52·6
   e) 10·3 + 25·8
   f) 51·8 + 20·7
   g) 39·3 + 19·8
   h) 16·3 + 17·8

3) Use Nuria's strategy to calculate:
   a) 5·7 + 3.2
   b) 6·3 + 8·5
   c) 3·1 + 12·5
   d) 50·4 + 6·7
   e) 30·9 + 5·8
   f) 70·7 + 2·4
   g) 15·4 + 12·2
   h) 33·5 + 22·2

4) Now calculate the following using a strategy of your choice.
   a) 8·8 + 3·9
   b) 4·7 + 2·6
   c) 3·7 + 4·5
   d) 22·3 + 15·8
   e) 31·4 + 25·9
   f) 36·7 + 21·8

CHALLENGE!

Copy and complete each pattern.

600 + 400 = ☐   60 + 40 = ☐   6 + 4 = ☐   0·6 + 0·4 = ☐

300 + ☐ = 1000   ☐ + 70 = 100   3 + 7 = ☐   ☐ + ☐ = 1

3.14 Adding and subtracting decimal fractions

We are learning to add and subtract decimal fractions using 'part-part-whole'.

Before we start

a) List all the different ways to make 6·9 by adding a whole number and a decimal fraction.

b) Now write two **decimal fractions** that total 6·9. How many ways can you find?

Number bonds help us to add and subtract decimal fractions.

Let's learn

Amman is solving the problem 22·6 + ◆ = 40

He knows that 6 tenths plus 4 tenths equals 10 tenths, or 1 whole.

He uses number bonds to help him solve the problem then draws an empty number line to check his answer.

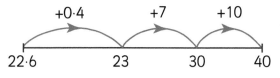

Add **4 tenths** to make 23 add **17** makes 40. The answer is **17·4**.

He draws a bar model and uses it to help him write two addition and two subtraction facts.

40	
22·6	17·4

22·6 + 17·4 = 40   17·4 + 22·6 = 40

40 − 22·6 = 17·4   40 − 17·4 = 22·6

1) Use number bonds to work out these calculations. Ask a friend to check your answers by drawing empty number lines.

   a) 14·3 + ♦ = 30   b) 37·8 + ♦ = 50   c) 71·5 + ♦ = 80

   d) 28·2 + ♦ = 50   e) 62·4 + ♦ = 80   f) 43·1 + ♦ = 90

   g) 16 = ♦ + 7·4   h) 25 = ♦ + 10·2   i) 42 = 21·8 + ♦

   j) ♦ + 52·4 = 100   k) ♦ + 36·7 = 100   l) ♦ + 85·6 = 100

2) Find the missing numbers. Write two addition facts and two subtraction facts for each bar model.

a)

?	
14·9	28·1

b)

71	
50·5	?

c)

90	
?	44·1

d)

?	
69·4	30·6

3) Amman calculates 17·4 − 6·1.

   17 minus 6 equals **11** and 0·4 minus 0·1 equals **0·3**. The answer is **11·3.**

   I can check my answer by adding.
   11·3 plus **0·1** equals 11·4
   11·4 plus **6** equals 17·4

   Calculate the answers to these subtractions. Check each answer by adding.

   a) 25·9 − 4·5   b) 50·7 − 9·3   c) 32·4 − 6·6

   d) 78·3 − 26·3   e) 60·4 − 47·2   f) 43·6 − 27

CHALLENGE!

Work with a partner. Write down four decimal fractions that add to 10. How many different ways can you find?

4.1 Recalling multiplication and division facts for 6

We are learning to recall multiplication and division facts for 6.

Before we start

Isla says that she can work out 12 × 2 using facts she already knows for 10 and 2. Explain how she can calculate 12 × 2 using facts for 10 and 2.

Recalling multiplication and division facts quickly helps us use strategies and solve problems efficiently.

Let's learn

Your knowledge of multiplication and division facts for 3 can help us recall facts for 6 quickly.

Let's look at 3 × 4

We can see that by doubling 3 × 4, we can work out the answer to 6 x 4

6 × 4 = 2 × (3 × 4)

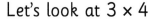

Let's practise

1) Show how you could use your knowledge of the 3 times tables to work out these problems involving 6. You could draw an array to help you.

   a) 6 × 2   b) 5 × 6   c) 6 × 6   d) 6 × 8

2) Match the questions with the correct answers.

   30   6   24   7   60   6   54   3   48

   a) 42 ÷ 6   b) 6 times 10   c) five lots of six

   d) 18 divided by 6   e) 1 × 6   f) 8 groups of 6

   g) 9 × 6   h) 36 split into 6

3) Work through these questions and think about which questions you recall quickly and which are more difficult to recall. This will help you know which facts you need to practise more.

a) $6 \times = 24$

b) $6 \times 9 =$

c) $7 \times 6 =$

d) $12 \div 6 =$

e) $ \times 6 = 60$

f) $36 \div 6 =$

g) $6 \times = 18$

h) $48 \div 6 =$

i) $6 \times = 30$

j) $6 \div 6 =$

CHALLENGE!

Work with a partner. Write down all the multiplication or division facts you want to practise on pieces of card or paper. Then write down all the answers to your facts on other pieces of card or paper.

Turn the two sets of cards over so you can't see them and shuffle them.

Now take turns to turn over a fact card and an answer card – do they match?

4 Number – multiplication and division

4.2 Recalling multiplication and division facts for 9

We are learning to recall multiplication and division facts for 9.

Before we start

Finlay is working out $32 \div 4$. How could he use his knowledge of multiplication facts for 4 to help him? Explain your reasoning.

Recalling multiplication and division facts quickly helps us use strategies and solve problems efficiently.

Let's learn

We can use our knowledge of the 10 times table to help us recall facts for 9 quickly.

Let's look at 3×9. If we know that $3 \times 10 = 30$ because we know multiplication facts for 10, then we can just subtract one group of three to get the answer to 3×9.

$3 \times 9 = (3 \times 10) - (3 \times 1)$

$\qquad = 30 - 3$

$\qquad = 27$

Let's practise

1) Match each question with the correct answer.

ten 9s

3 groups of 9

54 divided by 9

8 × 9

81 split into 9

18 shared into 9

five 9s

4 × 9

90

36

9

72

45

27

6

2

2) Look at a hundred square. Place one finger on the number 9 on the hundred square. Add 10 and subtract 1 by moving your finger down a row and back a square to the number 18.

Do this again and again, more and more quickly and say each multiple of 9 aloud each time.

Colour in the multiples of 9. Why do they make a diagonal pattern like this?

3) Choose four multiplication facts for 9 that you have to think about the most.

   a) Complete four multiplication triangles based on them.

   b) For each of your triangles, write a multiplication and a division fact.

Example

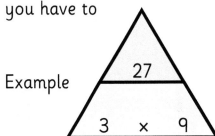

CHALLENGE!

Work with a partner. Write down the first five multiples of 9. Now add the digits of each multiple together, e.g. 2 × 9 = 18, 3 × 9 = 27.

What do you notice?

Now continue with the next five multiples. What do you notice this time?

Explore some more multiples of 9 that are even higher. Can you spot a rule that is always true?

How could this help you when you are recalling multiplication and division facts for 9?

4 Number – multiplication and division

4.3 Multiplying multiples of 10, 100 or 1000

We are learning to multiply whole numbers by multiples of 10, 100 or 1000.

Before we start

Amman tells Isla that 9500 divided by 10 is 95. Isla thinks the answer should be 950. Who is correct? Explain why.

We can use facts we know to help us solve multiplication problems involving multiples of 10, 100 and even 1000.

Let's learn

Let's look at 4 × 20. We know that there are 2 tens in 20.

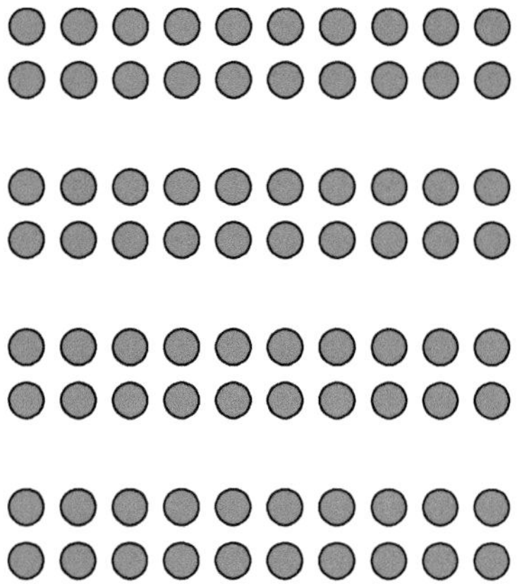

4 × 20   = 4 × 2 tens

          = 8 tens

          = 80

How about 4 × 200?

4 × 200   = 4 × 2 hundreds

          = 8 hundreds

          = 800

Or 4 × 2000?

4 × 2000   = 4 × 2 thousands

          = 8 thousands

          = 8000

If we know 4 × 2 we can use this fact to work out multiples of 10, 100 and 1000.

1) Match each question with a number fact you can use to help, then answer each of the questions.

40 × 4	6 × 3 = 18
900 × 5	6 × 5 = 30
800 × 3	4 × 4 = 16
70 × 2	9 × 2 = 18
500 × 5	7 × 2 = 14
60 × 3	8 × 4 = 32
800 × 4	8 × 3 = 24
90 × 2	5 × 5 = 25
60 × 5	9 × 5 = 45

2) Write three questions that you could use the fact 9 × 4 = 36 to answer.

3) The school is holding a raffle at the school concert. Can you work out how many raffle tickets have been sold altogether for each of these amounts sold?

a) Amman sold 90 strips of 5 tickets.
b) One class sold 300 strips of 5 tickets.
c) The total amount sold for the whole school was 2000 strips of 5 tickets.

CHALLENGE!

Isla is using a known fact to help work out a question. Her question has two zeroes in it. How many different questions could it be?

What about three zeroes?

I don't know the answer, but I do know what 6 × 6 is and I can use that to help.

4 Number – multiplication and division

4.4 Dividing by multiples of 10, 100 or 1000

We are learning to divide whole numbers by multiples of 10, 100 or 1000.

Before we start

You have 11 hundred pound notes, 12 ten pound notes and 26 pound coins. How much money do you have altogether?

We can use facts we know to help us solve division problems involving multiples of 10, 100 or 1000.

Let's learn

Let's look at 40 ÷ 2. We know there are 4 tens in 40.

We can use the fact **4 ÷ 2 = 2**

So 40 ÷ 2 = 4 tens ÷ 2

$\qquad$ = 2 tens

$\qquad$ = 20

What about 400 ÷ 2?

400 ÷ 2 = 4 hundreds ÷ 2

$\qquad$ = 2 hundreds

$\qquad$ = 200

If we know 4 ÷ 2 = 2 we can use this fact to work out multiples of 10, 100 and 1000.

Let's practise

1) For each of these division questions, write down a fact that you can use to help.

Then write the answer.

a) 240 ÷ 6

Known fact: 24 ÷ 6

So, 240 ÷ 6 = 40

b) 450 ÷ 5

Known fact:

So, 450 ÷ 5 =

c) 270 ÷ 3

Known fact:

So, 270 ÷ 3 =

d) 240 ÷ 4

Known fact:

So, 240 ÷ 4 =

2) Use the number facts below to help you answer each question on the balloon.

300 ÷ 6   150 ÷ 3   2400 ÷ 6   1200 ÷ 2   300 ÷ 3   1500 ÷ 5

15 ÷ 3 = 5   15 ÷ 5 = 3   24 ÷ 6 = 4
      12 ÷ 2 = 6   30 ÷ 6 = 5   30 ÷ 3 = 10

3) An order of 540 pencils arrived at school to be shared equally between the six classes. How many pencils will each class receive?

CHALLENGE!

You may be able to work out the answer to questions like 200 ÷ 5, but what if both numbers are multiples of 10 or a 100?

200 ÷ 50

Try to think of a fact you already know that might help you. Then test your method on these questions.

a) 120 ÷ 60   b) 150 ÷ 30   c) 210 ÷ 30

4.5 Solving division problems

We are learning to solve division problems by multiplying.

Before we start

3 × 6 = 18. Can you use this fact to help you work out 6 × 6? Explain how.

We can solve division problems by turning them into multiplication problems.

Let's learn

Let's work out 12 ÷ 3. To solve the problem, we can think of it as how many groups of 3 are in 12.

We could write it like this: × 3 = 12.

We know that **4** × 3 = 12, so we can use this fact to give the answer to 12 ÷ 3 = **4**.

We can **reverse** division problems and think of them as backwards multiplication questions. This is because division is the opposite action of multiplication. We call this the **inverse relationship** between multiplication and division.

Let's practise

1) Can you reverse these division problems and turn them into multiplication problems, then work out the answer using multiplication?

a) What is 10 divided by 2?   × 2 = 10 so 10 ÷ 2 =

b) What is 30 divided by 6?   × 6 = 30 so 30 ÷ 6 =

c) What is 45 divided by 9?   × 9   = 45 so 45 ÷ 9 =

2) There are 48 children at Games Club. They are split into equal teams. Use multiplying to work out how many teams there would be for:

a) teams of eight   b) teams of four   c) teams of three

Write the related multiplication and division fact for each one.

3) Answer each of these division questions by reversing them and thinking of them as multiplication problems.

Describe what you have done to work out the answer.

Example 55 pens are shared between five groups.

How many pens will each group have?

This is a division question ($55 \div 5 = ?$). I can reverse it and think about it as a multiplication question ($? \times 5 = 55$). I know that ten 5s make 50, so $11 \times 5 = 55$. So $55 \div 5 = 11$ pens each.

a) 40 coloured pencils are split between two groups. How many pencils will each group have?

b) 21 paint brushes are split between three groups. How many paintbrushes will each group have?

c) 40 sticks of glue are split between five groups. How many sticks of glue will each group have?

CHALLENGE!

Work with a partner. How many multiplication facts can you find that make the number 120?

Write them down and ask your partner to give the related division fact for each one.

Try thinking up other numbers to challenge each other.

4.6 Solving multiplication problems

We are learning to use partitioning to solve three-digit by one-digit multiplication problems.

Before we start

It costs £10 for a school pupil to go to the science museum. How much does this cost for a class of 32? Explain how you know.

We can split three-digit numbers by place value to make multiplication problems easier to solve.

Let's learn

Splitting up numbers into smaller numbers is called **partitioning.** We can use partitioning as a strategy to solve multiplication problems.

Let's look at 24 × 3.

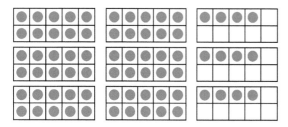

We can split 24 into tens and ones and multiply each by three. Then we add the answers together.

We can record this using brackets: $24 \times 3 = (20 \times 3) + (4 \times 3)$
$$= 60 + 12$$
$$= 72$$

We can also record this using a **grid.**

We can split three-digit numbers into hundreds, tens and ones.

Let's look at 124 × 3.

×	100	20	4
3	300	60	12
300 + 60 +12 = 372			

1) Answer the following using the grid method.

   a) 256 × 3

×	200	50	6
3	600	150	18

   600 + 150 + 18 = 768

   b) 498 × 5   c) 524 × 4   d) 375 × 6   e) 432 × 6   f) 674 × 8

2) Answer these using brackets.

   a) 634 × 4   b) 752 × 6   c) 372 × 5   d) 284 × 6   e) 378 × 4

3) A bookcase has nine shelves.

   How many books would there be altogether if each shelf had:
   a) 165 books   b) 352 books   c) 541 books
   Record your thinking using either the grid method or brackets.

CHALLENGE!

Work with a partner.

2	4	5	6

Arrange these four digit cards on the table like this.

×

How many different calculations can you make that are more than 1000 and less than 2000 by rearranging your four digit cards?

4 Number – multiplication and division

4.7 Using partitioning to solve division problems

We are learning to partition numbers to solve division problems.

Before we start

Nuria says that 90 × 5 gives the same answer as 9 × 50. Is she correct? Explain your answer.

We can use partitioning to make solving division problems easier.

Let's learn

Let's look at 96 ÷ 4.

We can split 96 into smaller numbers to make it easier. Think about numbers you know are divisible by 4. We know that 80 and 16 are both multiples of 4, so this would be easier than splitting 96 into 90 and 6.

We can divide 80 by 4, then 16 by 4 and add the answers together.

We can record this using brackets:

$$96 ÷ 4 = (80 ÷ 4) + (16 ÷ 4)$$
$$= 20 + 4$$
$$= 24$$

We could also record this using a grid:

÷	80	16
4	20	4
20 + 4 = 24		

Let's practise

1) Partition the larger number into two smaller sets that are both divisible by the smaller number.

Problem	Partition the larger number	Enter values in grid	Answer
64 ÷ 4	64 = 40 + 24	40 24 4 ☐ ☐	10 + 6 = 16
96 ÷ 8		☐ ☐	
84 ÷ 6		☐ ☐	

2) Partition the larger number into two smaller sets that are **both divisible by the smaller number**.

Problem	Partition the larger number	Enter the values in the grid	Answer
78 ÷ 3		☐ ☐	
84 ÷ 7		☐ ☐	
75 ÷ 5		☐ ☐	

3) There are a total number of 126 cupcakes at the party. How many plates would be needed if there were:

   a) six cakes on each plate
   b) nine cakes on each plate
   c) three cakes on each plate

   Record your thinking either using brackets or a grid.

CHALLENGE!

Work with a partner. You will need a set of numeral cards from 0 to 9. Shuffle the cards and place them face down.

Take two numeral cards to make a two-digit number. You can choose which way round you want to make the number.

Then your partner takes another numeral card. You need to divide the two-digit number by the second card you picked.

For example, if you picked 9 and 2 cards first, then your partner picked a 4, you would choose either to divide 92 by 4, or 29 divided by 4.

Think about how you could partition the numbers to make it easier.

Some of your answers will have a remainder.

4.8 Using rounding and compensating to solve multiplication problems

We are learning to use rounding and compensating to solve multiplication problems.

Before we start

Amman plants four rows of potatoes, with 14 potatoes in each row. Work out how many potatoes he plants altogether. Round your answer to the nearest 10.

We can use rounding and compensating as a strategy to make multiplication problems easier.

Let's learn

Let's look at 19 × 3.

We could **round** the problem to 20 × 3 first because this is easier than working out 19 × 3.

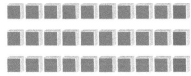

We know 20 × 3 = 60. Then we need to **compensate** by taking 1 cube from each row of 20 to make it back into 19 × 3.

So, 19 × 3 = (20 × 3) − (1 × 3)

$\qquad$ = 60 − 3

$\qquad$ = 57

We could also record this using an empty number line.

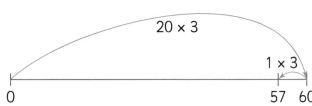

We could also work out 21 × 3 using the same strategy.

This time we **round** the problem to 20 × 3, then we need to **compensate** by adding on one more cube to each row to make it back into 21 × 3.

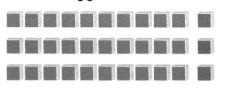

So, 21 × 3 = (20 × 3) + (1 × 3)

$\qquad$ = 60 + 3

$\qquad$ = 63

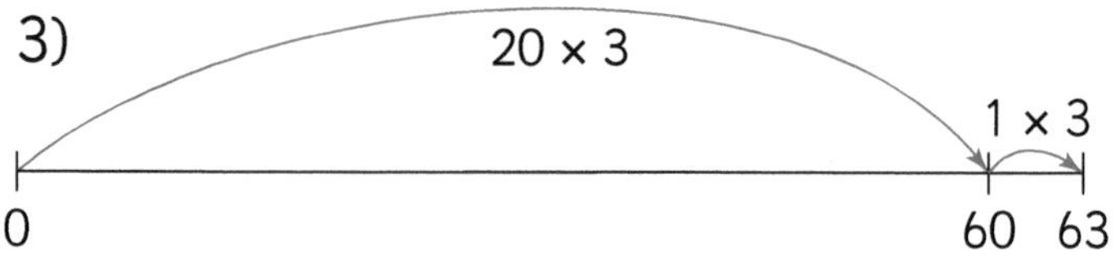

20 × 3

1 × 3

0   60   63

We could also record this using an empty number line.

Let's practise

1) Each of these calculations involves multiplying by nine.

Use a set of 10 frames to make groups of 10. Then take one from each group to compensate.

For example: 8 × 9

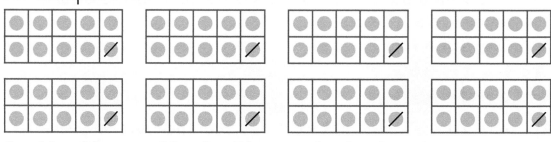

8 × 10 = 80   80 − 8 = 72   So, 8 × 9 = 72

a) 5 × 9

$\qquad$ 5 × 10 =   −   =   So, 5 × 9 =

b) 7 × 9

$\qquad$ 7 × 10 =   −   =   So, 7 × 9 =

c) 4×9

$4 \times 10 =$ $-$ $=$ So, $4 \times 9 =$

d) 6×9

$6 \times 10 =$ $-$ $=$ So, $6 \times 9 =$

2) If you know that $6 \times 30 = 180$, use rounding and compensating to work out the answer to:

a) 6×29 b) 6×31 c) 6×28

3) It costs £48 for a ticket to the concert. How much will it cost altogether for each of these people?

a) Sam buys two tickets.
b) Calum buys five tickets.
c) Molly buys six tickets.
d) Kirsty buys nine tickets.

Record your strategy for solving each problem.

CHALLENGE!

Finlay multiplied a number by 21. If his answer was between 100 and 200, can you work out all the numbers he might have multiplied by 21?

4.9 Solving problems involving addition, subtraction, multiplication and division

We are learning to solve problems that involve a combination of addition, subtraction, multiplication and division.

Before we start

Finlay is working out 16 × 5. Can you think of three different ways you could solve this problem? Explain which strategy you found the most efficient and why.

When we are solving problems we often need to use a combination of addition, subtraction, multiplication and division.

Let's learn

James has £14 in his bank account. He starts a new job earning the same amount every week and puts this straight into his bank account. Four weeks later he has £46. How much did he earn each week?

Think about what we need to work out.

First, we need to know how much he has earned altogether.

Then think how we can calculate this. We can do this by subtracting the amount he had at the start from the amount he had after four weeks.

46 – 14 = 32

Then we need to divide this by four to work out how much he was paid each week.

32 ÷ 4 = 8

James was paid £8 each week.

Think carefully about the calculations you need to work out to solve these problems, and record the strategies you use for each step.

1) Isla and Finlay were picking raspberries. Isla picked four crates of raspberries. Finlay picked six crates of raspberries. Each crate contained eight punnets.

   a) How many punnets did Isla pick?
   b) How many more punnets did Finlay pick?

2) The school needed some new art supplies. Can you work out how much each of these orders will cost altogether?

   a) 20 new paint brushes, which cost £2 for each pack of five.
   b) 50 glue spreaders, which cost £3 for each pack of 10.
   c) 15 tubes of blue paint and six tubes of red paint. These cost £4 for a pack of three.
   d) 24 paint palettes at £3 for each pack of four.
   e) 200 pencils. These cost £5 for a pack containing 50 pencils.
   f) What will the total amount of all the art supplies cost?

3) At the theme park it costs £11 for an adult day ticket and £6 for a child day ticket. A family ticket costs £30, and is valid for up to three children.

   a) How much does it cost for one adult and three children?
   b) How much does it cost for two adults and two children?
   c) What would be the difference for both families if they bought a family ticket?

CHALLENGE!

Work with a partner. Make up a problem for each other that would involve at least two different calculations to solve.

Record the strategies you used and share your thinking with each other.

4 Number – multiplication and division

4.10 Multiplying decimal fractions

We are learning to multiply decimal fractions involving tenths by 10, 100 and 1000.

Before we start

Isla has solved the first problem but is finding the rest of the problems tricky. Explain how you would help her.

a) $4 \times 5 = 20$

b) $40 \times 5 =$

c) $4 \times 500 =$

d) $4000 \times 5 =$

We know that when we multiply a number by 10, it gets 10 times bigger.

Let's learn

When we multiply a number by 10, the digits move one place to the left and a zero is written in the empty column as a placeholder.

When we multiply a number by 100, the digits move two places to the left, with zero as a placeholder.

When we multiply a number by 1000, the digits move three places to the left, with zero as a placeholder.

The same is true when multiplying decimal fractions.

	THOUSANDS	HUNDREDS	TENS	ONES		TENTHS
				0		4
0.4×10				4		0
0.4×100			4	0		0
0.4×1000		4	0	0		0

$0.4 \times 10 = 4$

$0.4 \times 100 = 40$

$0.4 \times 1000 = 400$

Remember that the digits move and not the decimal point. The decimal point is always fixed between the ones and the tenths.

Let's practise

1) Multiply decimals by 10 using the place value grids below, or use your own decimats or decimal sliders.

Tens	Ones	Decimal point	Tenths
	0	.	2

× 10

Tens	Ones	Decimal point	Tenths
	2	.	0

a)

Tens	Ones	Decimal point	Tenths
	0	.	6

× 10

Tens	Ones	Decimal point	Tenths
		.	

b)

Tens	Ones	Decimal point	Tenths
	0	.	8

× 10

Tens	Ones	Decimal point	Tenths
		.	

c)

Tens	Ones	Decimal point	Tenths
	0	.	9

× 10

Tens	Ones	Decimal point	Tenths
		.	

2) Work out the answers to these:

a) 5.7×10 b) 6.2×100 c) 8.8×1000

d) 12.2×100 e) $14.2 \times \quad = 142$ f) $2.8 \times \quad = 2800$

3) There are 100 cm in a metre. Can you convert these measurements from metres to centimetres?

a) 3·2 m b) 17·9 m c) 0·8 m d) 10·1 m

CHALLENGE!

Nuria thinks of a number. She divides it by 100. Then she adds 20. Then she divides it by 10. Her answer is 2·4. What number did she start with? How do you know? Can you make up more problems like this for a partner?

4 Number – multiplication and division

4.11 Dividing whole numbers by 10, 100 and 1000

We are learning to divide whole numbers by 10, 100 and 1000 with answers involving tenths.

Before we start

Nuria works out that $1800 ÷ 10 = 180$. Is she correct? Explain how you know.

We know that when we divide by 10, the number gets 10 times smaller, and when we divide by 100, the number gets 100 times smaller. The same is true when the answer involves tenths.

Let's learn

When a number is divided by 10, the digits move one place to the right.

THOUSANDS	HUNDREDS	TENS	ONES	TENTHS
		2	3	0
			2	3

$23 ÷ 10 = 2·3$

When a number is divided by 100, the digits move two places to the right, and when a number is divided by 1000 the digits move three places to the right.

THOUSANDS	HUNDREDS	TENS	ONES	TENTHS
	2	3	0	0
			2	3

$230 ÷ 100 = 2·3$

Remember that the digits move and not the decimal point. The decimal point is always fixed between the ones and the tenths.

THOUSANDS	HUNDREDS	TENS	ONES	TENTHS
2	3	0	0	0
			2	3

$2300 ÷ 1000 = 2·3$

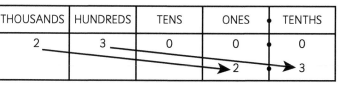

Let's practise

1) Divide these whole numbers by 10. You could use a place value grid to help you.

a) 17   b) 21   c) 48   d) 66

2) Divide these numbers by 10:

a) 46   b) 83   c) 427   d) 903

Divide these numbers by 100:

e) 80   f) 660

Divide these numbers by 1000:

g) 8100   h) 900

3) There are 1000 millilitres in each litre. Can you convert these measurements from millilitres to litres?

a) 200 ml   b) 1800 ml   c) 3000 ml   d) 5900 ml

CHALLENGE!

Isla thinks of a number. She multiplies it by 100. Then she adds 4. Then she multiplies it by 100. The answer is 8400. What number did she start with? How do you know?

Can you make up more problems like this for a partner?

4 Number – multiplication and division

4.12 Solving division problems with remainders

We are learning to solve division problems with remainders.

Before we start

Can you put these numbers in order from smallest to biggest?

4·19   0·18 $\frac{1}{2}$ 4·2 $5\frac{1}{2}$ 4·31

When we solve division problems that have remainders, the way we write the remainder depends on what the problem was about.

Let's learn

Sometimes we write remainders as whole numbers.

Amy's hens lay 23 eggs. She packs them into egg boxes of six. How many boxes does she need?

23 ÷ 6 = 3 r 5

Amy will need an extra box for the remainder, so will need four boxes altogether. For this problem it is most useful to write the remainder as a whole number to see how many boxes she will need.

Jack and five friends went to the cinema. It cost £33 altogether. How much did each ticket cost?

33 ÷ 6 = 5 r 5

It would be more useful to record this as £5.50. When we are working out problems involving money, the remainder is most useful written as a decimal.

It takes Sarah 30 minutes to run four laps around the track. How long does it take her to run one lap?

30 ÷ 4 = 7 r 2

It would be more useful to record this result as $7\frac{1}{2}$ minutes. Sometimes it is useful to write the remainder as a fraction.

Let's practise

1) There are 32 people in a class. The teacher puts them into teams of five to play football.

   a) How many teams can be made?

   b) How would you write the remainder for this problem? Explain why.

2) Think about the best way to write the remainders for these problems. Choose the answer you think works best from this box:

   $6 \cdot 5$ $6\frac{1}{2}$ $6 \cdot 50$ $6 \text{ r } 3$

   a) Four friends go to lunch. The bill comes to £26. How much should they each pay if they split the bill equally?

   b) Molly ran 39 kilometres over six days. How many km did she run on average each day?

   c) Ben has 26 biscuits. If he shares them between himself and three friends equally, how many biscuits would each of them get?

   d) There are 26 people at the birthday dinner. Tables seat four. How many tables will be needed?

   Explain why you picked each answer.

3) Jamie earns £12 every week for his paper round. How many weeks will it take for him to save up for these amounts? How will you write the remainder this time?

   a) £30   b) £59   c) £100

⭐ **CHALLENGE!** ⋯⋯⋯⋯⋯⋯⋯⋯⋯⋯⋯⋯⋯⋯⋯⋯⋯⋯⋯⋯⋯⋯⋯⋯⋯⋯

Can you come up with a word problem for each of these division sentences? Think about how the remainder is written when you come up with your problem.

$45 \div 10 = 4 \cdot 5$ $48 \div 5 = 9 \text{ r } 3$ $28 \div 8 = 3\frac{1}{2}$ $33 \div 4 = 8.25$

4 Number – multiplication and division

4.13 Solving multiplication and division problems

We are learning to use a range of strategies to solve multiplication and division problems.

Before we start

Finlay makes 48 muffins. He puts 12 muffins in each cake tin. Work out how many tins he needs and explain how you got your answer. Could you think of a different way to solve this problem?

When we have a range of strategies to solve multiplication and division problems, we can choose the strategy that is best suited to a particular problem.

Let's learn

Let's look at 4 × 28. We could use a range of strategies to solve this problem.

We could partition the numbers using place value like this:

4 × 28 = (4 × 20) + (4 × 8)

          = 80 + 32

          = 112

Or we could use rounding and compensating:

Or we could double and double again:

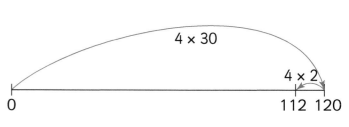

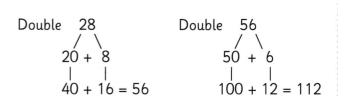

What about 54 ÷ 3?

We could partition the numbers like this:

$$54 ÷ 3 = (30 ÷ 3) + (24 ÷ 3)$$
$$= 10 + 8$$
$$= 18$$

Or we could use multiplication: $3 × \underline{\quad} = 54$

I know that $20 × 3 = 60$, and $60 - 54 = 6$ which is $2 × 3$.

$20 - 2 = 18$, so the answer must be 18.

You may have a different strategy! Use whatever strategy you think works best for these problems and remember to record your thinking.

Let's practise

1) Solve the following word problems using a multiplication strategy of your choice.

   a) Six coaches, each carrying 58 people, are going to the theatre. How many people in total are going?

   b) An average of 702 people visit a shop every hour. How many people will have visited the shop in eight hours?

   c) The ingredients for a cake have a mass of 611 g. What is the total mass of ingredients needed to make four cakes?

2) Three furniture companies need to ship tables to different locations. All the companies use trucks that can transport a maximum of 55 tables. Which of the companies will NOT be able to fulfil the shipment?

	Fast Track	Safe Load	Quick Star
Number of tables	540	220	275
Number of locations	9	4	5

4

3) Play a game, Number Boats, with a partner. Your aim is to sail your boat across a sea of numbers to the shore (Finish). Play in turn.

- Place a counter (your boat) on the Start square.
- Spin a 2–9 spinner and move your counter that number of squares.
- Multiply the number you land on by the number spun.
- If you multiply correctly, remain on the square. If you are incorrect you must move your counter back to the square you were on previously.
- If you land on a red number, you must explain your multiplication strategy to your opponent.
- Landing on a propeller square causes a breakdown in power and you must miss a turn.

401	🌀	212	698	470	319
359	139	🌀	422	579	182
890	222	999	777	311	🌀
665	802	🌀	Finish	🌀	548
🌀	681	888	709	949	721
413	🌀	499	218	532	303
561	365	🌀	732	289	Start

⭐ **CHALLENGE!**

Work with a partner. Make a Think Board like this:

Can you use four different strategies to solve this problem and record them on each space on the Think Board? You can use pictures or diagrams to help show your thinking.

Put a smiley face or a thumbs up on the strategy you think was the most efficient.

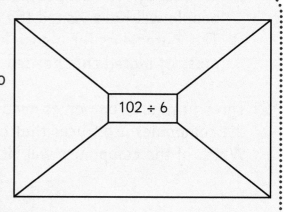

$102 \div 6$

5 Multiples, factors and primes

5.1 Identifying all factors of a number

We are learning to find all factors of a number.

Before we start

Amman has £6. Nuria has four times as much money as Amman. Can you write a multiplication sentence to show how much money Nuria has?

We can find all the factors of a number by investigating which whole numbers can divide exactly into the number.

Let's learn

Let's look at 12. It's helpful to start investigating numbers from 1 to 9. We can take each number and check if it will divide exactly into 12.

1) Does 1 divide exactly into 12? Yes, 12 ÷ 1 = 12. So we know that **1** and **12** are factors of 12.

2) Does 2 divide exactly into 12? Yes, 12 ÷ 2 = 6. So we know that **2** and **6** are factors of 12.

3) Does 3 divide exactly into 12? Yes, 12 ÷ 3 = 4. So we know that **3** and **4** are factors of 12.

4) We already know 4 divides into 12 from investigating 3.

5) Does 5 divide exactly into 12? No. So we know that 5 is NOT a factor of 12.

6) We already know 6 divides into 12 from investigating 2.

7) Does 7 divide exactly into 12? No. So we know that 7 is NOT a factor of 12.

8) Does 8 divide exactly into 12? No. So we know that 8 is NOT a factor of 12.

9) Does 9 divide exactly into 12? No. So we know that 9 is NOT a factor of 12.

We can also list the multiplication facts that make 12

$1 \times 12 = 12$

$2 \times 6 = 12$

$3 \times 4 = 12$

So 1, 2, 3, 4, 6 and 12 are factors of 12

Let's practise

1) Can you find all the factors of these numbers?

   a) 6 　　　　　　　　b) 10 　　　　　　　　c) 25

2) Look at this Bingo Board:

   Find all the factors of 24. Cross them out.
   Now find all the factors of 35 and cross them out.
   Have you won Bingo or are there any numbers left?

8	5	3
4	2	9
1	6	7

3) How many factors do these numbers have?

   a) 2 　　　b) 15 　　　c) 16 　　　d) 64

CHALLENGE!

Amman is thinking of a number. It's a two-digit number and it is between 50 and 70. It has 12 factors. Can you find the number he is thinking about?

5 Multiples, factors and primes

5.2 Identifying multiples of numbers

We are learning to find multiples of numbers.

Before we start

Which of these numbers are multiples of 10?

Write a multiplication or division fact for each multiple you have identified.

| 20 | 45 | 66 |

| 70 | 95 |

Multiples are made by multiplying a number by 1, 2, 3, 4… and so on.

Let's learn

Knowing about multiples of numbers can help us solve multiplication and division problems.

For example, let's look at multiples of 5:

The first twelve multiples of 5 are 5, 10, 15, 20, 25, 30, 35, 40, 45, 50, 55, 60.

What do you notice about all these multiples? Can you see a pattern? Would this continue if we carried on listing multiples?

We can see that all the multiples of 5 end in either 5 or 0.

If we were working out a problem like 165 ÷ 5 we would then know that our answer should not have a remainder because it ends in 5.

5

1) Which of these numbers are not a multiple of 4? How do you know?

   8   23   16   51   48   65   114   54

2) Write down the first 10 multiples of:

   a) 3   b) 6   c) 9

   What do you notice about each sequence of numbers?

3) Are these statements true or false? Circle true or false.

63 is a multiple of 9.	**True/False**
The fourth multiple of 2 is 12.	**True/False**
The multiples of 7 are all odd numbers.	**True/False**
40 is not a multiple of 10.	**True/False**
12 is a multiple of 3 and a multiple of 4.	**True/False**
The tenth multiple of 10 is 100.	**True/False**
Multiples of 8 are always even.	**True/False**
6 is a multiple of 12.	**True/False**

4) On a hundred square, colour in all the multiples of:

   a) 8   b) 6   c) 4   d) 2

   What do you notice about the patterns you have made?

CHALLENGE!

Write down all the multiples of 2. Now write down all the multiples of 3. Circle all the numbers that are multiples of both 2 and 3. The smallest of these numbers is called the lowest common multiple.

Can you do the same for 3 and 5? What is the smallest multiple these numbers have in common?

What about 2 and 5?

6 Fractions, decimal fractions and percentages

6.1 Identifying equivalent fractions

Before we start

Find two fractions that are equivalent to one third:

one third

We are learning to identify and create equivalent fractions.

We can convert fractions into tenths and hundredths using what we know about equivalence and simplification.

Let's learn

We know that we can find fractions that are equal by splitting them into smaller parts. Instead of getting one half of a bar, we might split the half into two equal parts and get two quarters:

one half

two quarters

If we want to write one half as a decimal, then changing it into quarters doesn't help us. Changing one half into three sixths or four eighths doesn't help us either:

three sixths

four eighths

Decimal fractions can only be written using tenths or hundredths (or thousandths, etc). So we must find an equivalent fraction in tenths or hundredths:

five tenths ✓
0·5

50 hundredths ✓
0·50

Let's practise

1) Finlay is trying to change a quarter into tenths:

one quarter

two eighths

three twelfths

If we split a quarter into two equal parts, we create two eighths. If we split a quarter into three equal parts, we create three twelfths. I've proven that it can't be done.

Which of the following fractions can be changed into tenths? Can you prove it?

a)

one third

b)

two quarters

c)

one fifth

d)

three fifths

e)

one sixth

f)

three eighths

2) Nuria has been asked to change one quarter into hundredths:

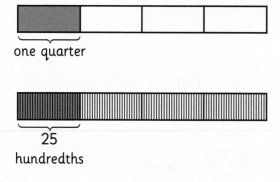

one quarter

25 hundredths

If we split each quarter into 25 parts there are 100 parts altogether. We have created hundredths. So one quarter is equivalent to 25 hundredths.

Which of the following fractions can be changed into hundredths? Can you prove it?

a)

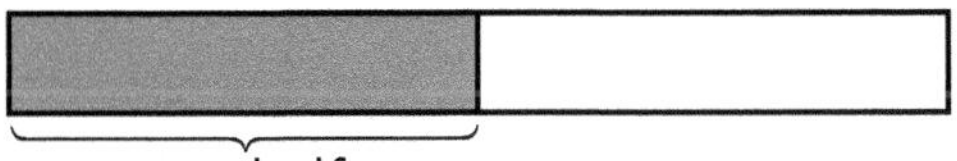

one half

b)

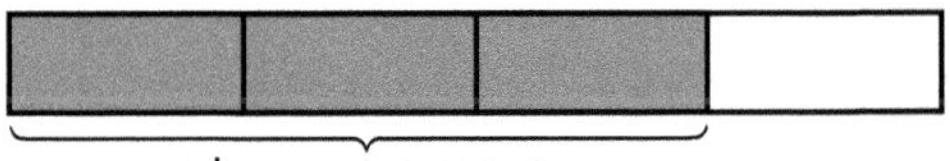

three quarters

c)

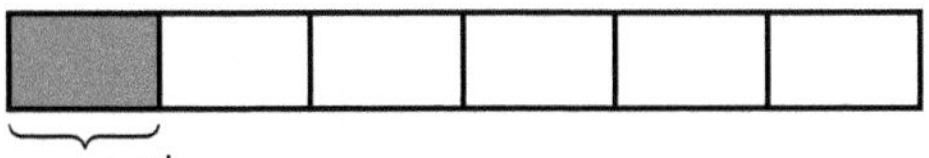

one sixth

d)

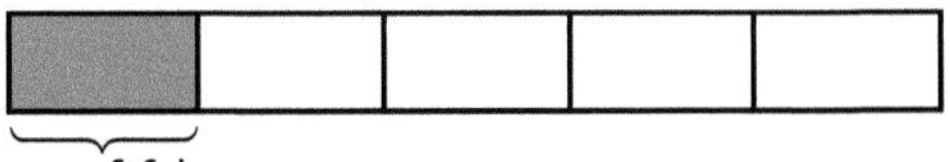

one fifth

e)

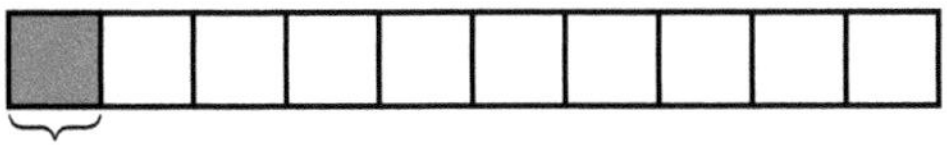

one
tenth

f)

one third

CHALLENGE!

Isla and Amman have been asked to find other fractions that have an equivalent in tenths:

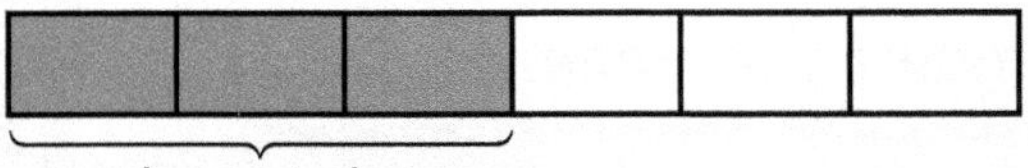

three sixths

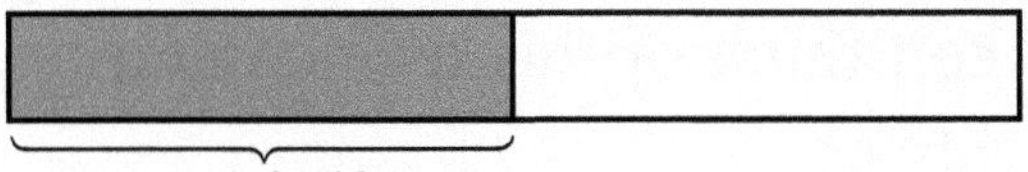

one half

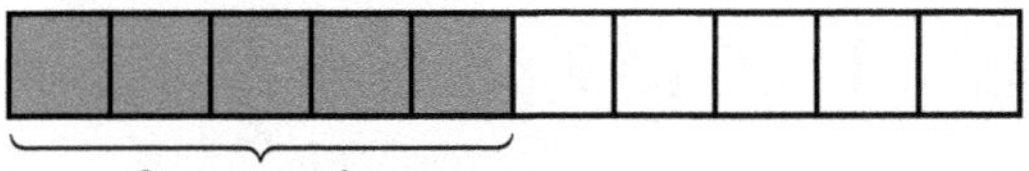

five tenths

Three sixths doesn't have an equivalent fraction using tenths because we can't split sixths into equal parts to make 10 altogether.

What if we **simplified** three sixths first? Three sixths is equal to one half. We can convert one half into five tenths.

Use simplification and/or equivalence to find which of the following can be converted into tenths:

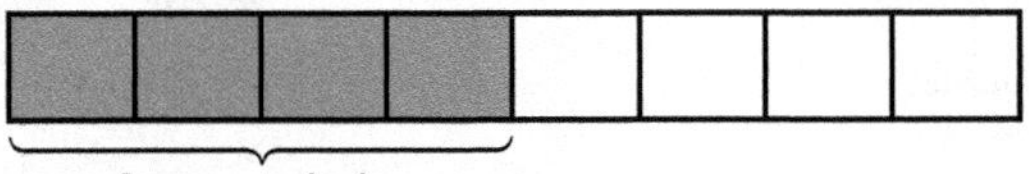

four eighths

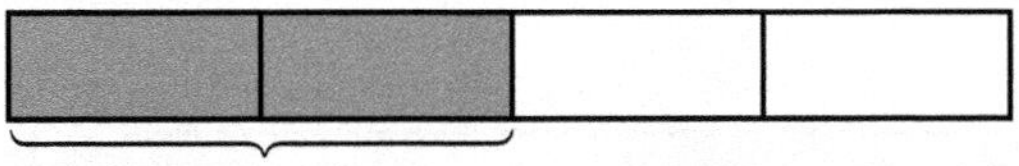

two quarters

one third

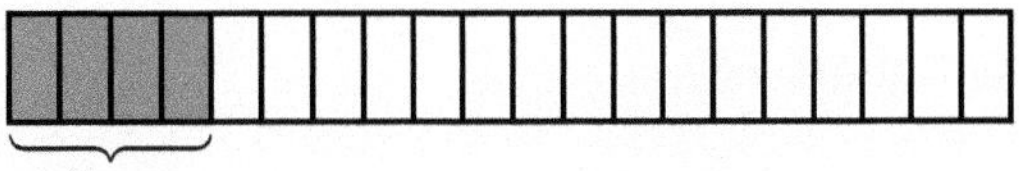

four twentieths

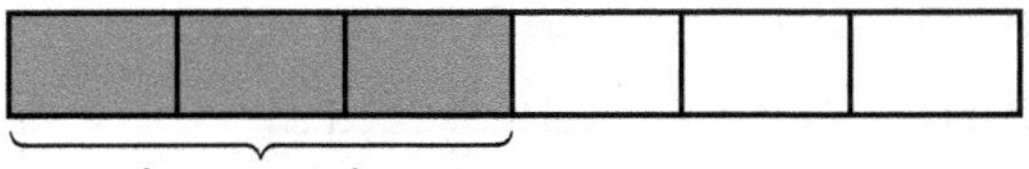

three sixths

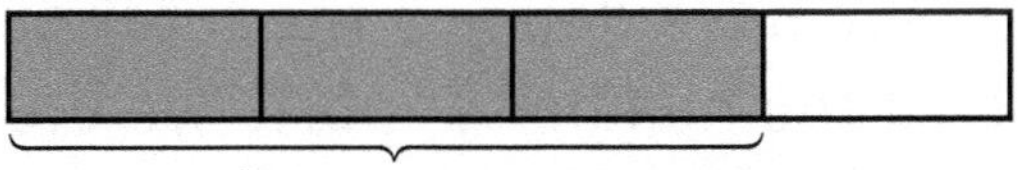

three quarters

6 Fractions, decimal fractions and percentages

6.2 Calculating equivalent fractions

We are learning to calculate equivalent fractions.

Before we start

Find two fractions that are equivalent to five sixths:

five sixths

We can multiply the number of parts in a fraction to find an equivalent. The amount we are getting remains the same.

Let's learn

If we are to get one half of a bar we can think of this as meaning 'one *out of* two parts'. We can write this as $\frac{1}{2}$:

one half

$=$ **1** *out of* **2** parts $= \dfrac{1}{2}$

To find an equivalent fraction we can split the halves into smaller parts. Two halves split into two parts creates four quarters:

The **2** tells us how many parts the whole has been split into and the **1** tells us how many of those parts we are getting.

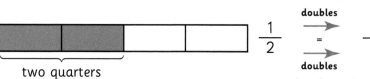

two quarters

number of parts we are getting

doubles

$\dfrac{1}{2}$ $\xrightarrow{\ \ \ \ \ }$ $=$ $\xrightarrow{\ \ \ \ \ }$ $\dfrac{2}{4}$

doubles

number of parts that make up the 'whole'

Instead of getting **1** *out of* **2** parts, we are now getting **2** *out of* **4** parts. We have doubled the total number of parts, so we have also doubled the number of parts we are getting. We are still getting exactly the same amount of the bar.

Let's practise

1) Use Finlay's method of doubling to find equivalent fractions for the following – the first one has been done for you:

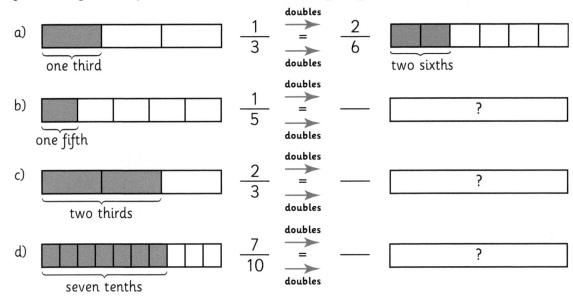

a) one third $\dfrac{1}{3}$ —doubles→ = —doubles→ $\dfrac{2}{6}$ two sixths

b) one fifth $\dfrac{1}{5}$ —doubles→ = —doubles→ ___   ?

c) two thirds $\dfrac{2}{3}$ —doubles→ = —doubles→ ___   ?

d) seven tenths $\dfrac{7}{10}$ —doubles→ = —doubles→ ___   ?

2) Nuria finds more equivalents by multiplying the number of parts by three, four and five. Use Nuria's strategy to find three equivalent fractions for each of the fractions in question 1. (Nuria has done the first one for you.)

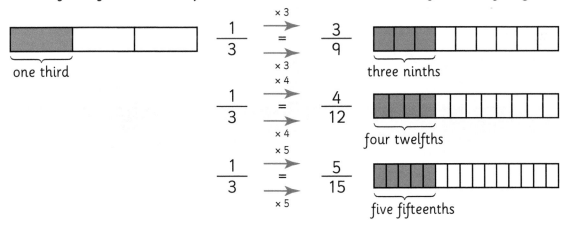

one third

$\dfrac{1}{3}$ —×3→ = —×3→ $\dfrac{3}{9}$ three ninths

$\dfrac{1}{3}$ —×4→ = —×4→ $\dfrac{4}{12}$ four twelfths

$\dfrac{1}{3}$ —×5→ = —×5→ $\dfrac{5}{15}$ five fifteenths

CHALLENGE!

Find as many equivalent fractions as you can that are the same as:

one half   three fifths

6 Fractions, decimal fractions and percentages

6.3 Comparing and ordering fractions

We are learning to compare and order fractions using equivalence.

Before we start

Place these fractions in order from smallest to largest: $\frac{1}{2}$ $\frac{1}{5}$ $\frac{1}{10}$ $\frac{1}{4}$ $\frac{1}{3}$

Equivalence helps us compare and order fractions that are not alike.

Let's learn

Would you rather get five eighths or three quarters of this bar of chocolate?

Five eighths sounds like more but I'm not sure how to work this out?

Eighths and quarters are different sizes, so it is difficult to compare them as they are:

three quarters

five eighths

We can use equivalence to help us out:

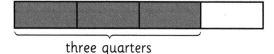

three quarters

$$\frac{3}{4} \xrightarrow[\times 2]{\times 2} = \frac{6}{8}$$

six eighths

Six eighths is bigger than five eighths, so we can say that Nuria will get more chocolate if she chooses three quarters rather than five eighths.

We can say: $\frac{3}{4}$ **is greater than** $\frac{5}{8}$ of the bar.

6.3

Let's practise

1) Use equivalence to help the children choose the largest portion of chocolate:

a) 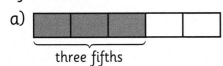   or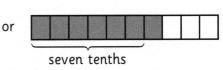

three fifths   seven tenths

b) 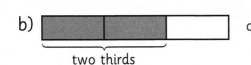   or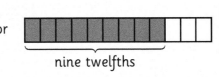

two thirds   nine twelfths

c) 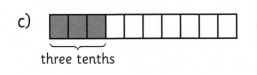   or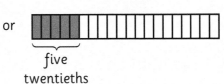

three tenths   five
twentieths

2) Use equivalence to find if the following fractions are smaller or larger than one half (the first one has been done for you):

$\dfrac{6}{10}$ $\dfrac{5}{12}$ $\dfrac{11}{20}$ $\dfrac{5}{8}$ $\dfrac{30}{50}$ $\dfrac{45}{100}$

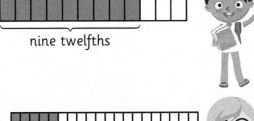

$\dfrac{1}{2} \xrightarrow[\times 5]{\times 5} = \dfrac{5}{10}$ So $\dfrac{6}{10}$ is larger than $\dfrac{1}{2}$

3) a) Use equivalence to order these fractions from smallest to largest:

$\dfrac{3}{5}$ $\dfrac{3}{4}$ $\dfrac{7}{10}$ $\dfrac{13}{20}$ $\dfrac{1}{2}$

b) Copy the number line below and write each fraction in the correct position:

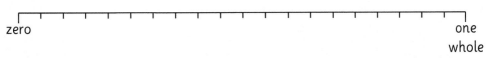

zero   one
whole

CHALLENGE!

a) Amman ate one quarter of a cake. He cut what was left into six equal slices and ate three slices.

Finlay had an identical cake. He ate one half of the cake. He cut what was left into four slices and ate two slices. Who ate more cake?

b) Nuria and Isla each have a one litre bottle of orange juice. Nuria drinks one third of her bottle. She then shares the rest between four cups and drinks three of them. Isla drinks one quarter of her bottle. She shares the rest between nine cups and drinks six of them. Who drinks more orange juice?

Fractions, decimal fractions and percentages

6.4 Decimal equivalents to tenths and hundredths

We are learning to recognise and write decimal equivalents of any number of tenths and hundredths.

Before we start

Identify which of the following fractions can be converted into tenths or hundredths:

$\frac{3}{4}$　　$\frac{4}{5}$　　$\frac{6}{20}$　　$\frac{2}{3}$　　$\frac{15}{20}$　　$\frac{5}{8}$

One hundredth is 10 times smaller than one tenth.

Let's learn

We know that the first place after the decimal point represents tenths.

2 tenths = 0·2　　5 tenths = 0·5

10 tenths = 1·0 (because 10 tenths makes 1 whole)

One hundredth is 10 times smaller than one tenth so the second place after the decimal point represents hundredths.

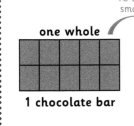

10 times smaller　　10 times smaller

one whole　　**one tenth**　　**one hundredth**

1 chocolate bar　　0·1 chocolate bars　　0·01 chocolate bars

One hundredth is 10 times smaller than one tenth. We can write this as 0·01.

one hundredth = 0·01

 13 hundredths　=　 **1 tenth**　+　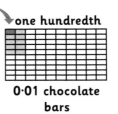 **3 hundredths**

So, we can write
**13 hundredths
= 0·13**

1) Write each of the portions below as a decimal fraction.

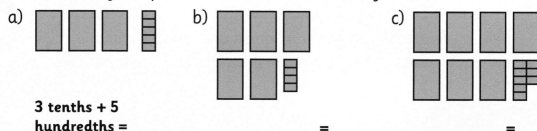

a)

3 tenths + 5 hundredths = _____

b)   = _____

c)   = _____

2) Write the amount of chocolate for each diagram as a fraction and a decimal fraction:

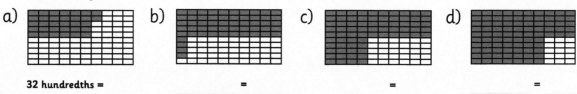

a)   b)   c)   d)

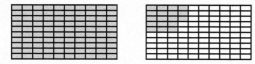

32 hundredths = _____   _____ · _____   _____ = _____   _____ = _____

3) Isla has this amount of cake:

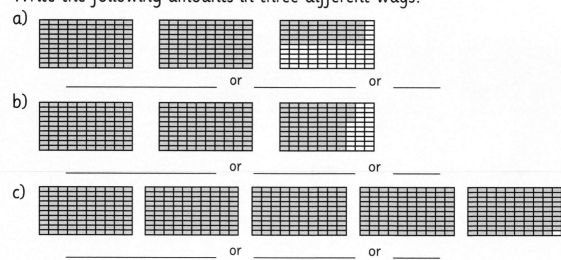

She can write this as:

1 whole and eighteen hundredths or **118 hundredths** or **1·18**

Write the following amounts in three different ways:

a)

_____ or _____ or _____

b)

_____ or _____ or _____

c)

_____ or _____ or _____

CHALLENGE!

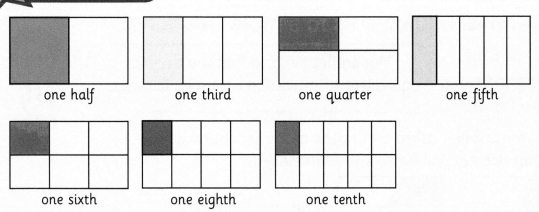

one half   one third   one quarter   one fifth

one sixth   one eighth   one tenth

Nuria has been given the fraction cards above and has to convert each into a decimal fraction to two decimal places.

Do you agree with Nuria's statement below?

Write the decimal fractions for the fractions that can be converted into tenths or hundredths.

> Some of these can't be converted into tenths or hundredths so I can't write the decimal fraction.

6.5 Decimal equivalents to simple fractions

We are learning to find the decimal equivalent of a fraction (using equivalence).

Before we start

Amman has been asked to find as many fractions as he can that are equivalent to the following:

 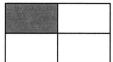

Can you help him?

Equivalence can help us convert simple fractions to decimal fractions.

Let's learn

We know that the first place after the decimal point is tenths and the second place after the decimal point is hundredths. Each decimal place to the right gets 10 times smaller.

one tenth

0·1 chocolate bars

one hundredth

0·01 chocolate bars

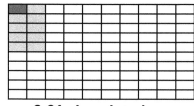

How do we write other fractions like halves, quarters and fifths as a decimal fraction? Is one half the same as 0·1?

No, 0·1 equals one tenth. If we want to write these as decimal fractions we must find their equivalent fractions as tenths or hundredths.

If we want to write one half as a decimal fraction, we need to convert it into tenths first, so:

 = 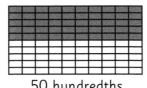 = 0·5

one half 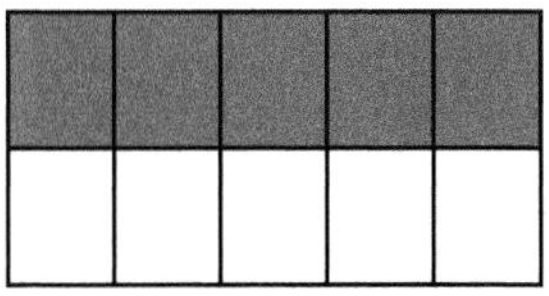   five tenths

One half can also be converted into hundredths, so:

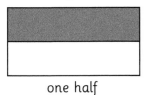 = 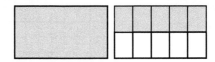 = 0·50

one half   50 hundredths

Let's practise

1) Write each portion as a decimal fraction:

a)

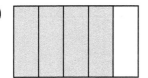

    four fifths   = _____   = _____

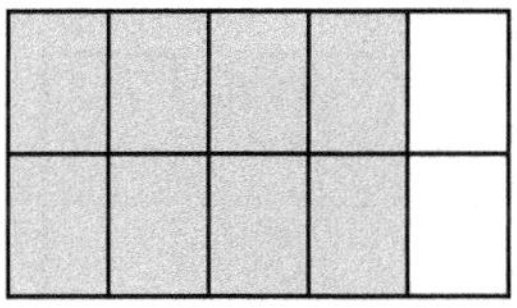

b)

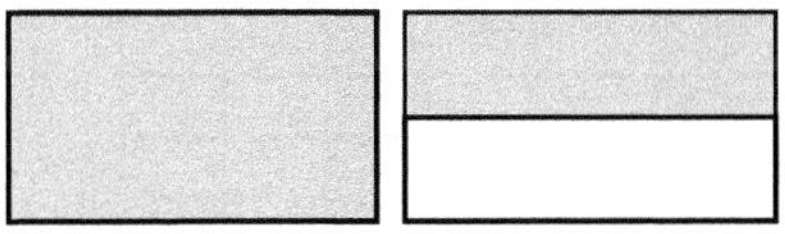

    one and a half   = _____   = _____

c)

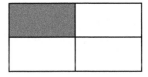

    one quarter   = _____   = _____

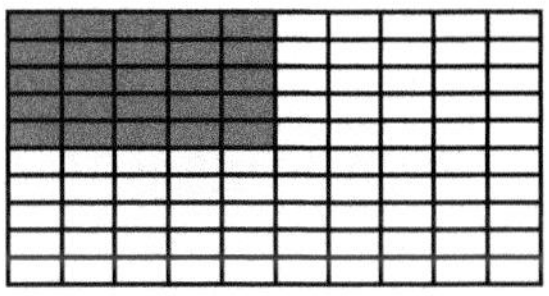

2) Change each fraction into tenths or hundredths and write them as a decimal fraction:

a)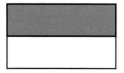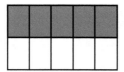

one half   = <u>5 tenths</u> = 0·5

b)

three quarters = _____ = ___

c)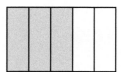

three fifths   = _____ = ___

d)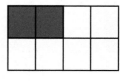

two eighths   = _____ = ___

CHALLENGE!

Nuria has been given more fraction cards and has to convert each into a decimal fraction:

three sixths

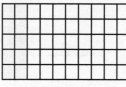

six eighths

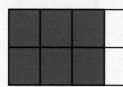

three twelfths

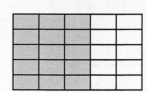

14 twentieths

10 fiftieths

15 twenty-fifths

I'm not sure how to convert some of these fractions into tenths or hundredths. Can you help?

Help Nuria find a way of writing each fraction as a decimal fraction.

Hint: It may help to simplify some of the fractions first.

6 Fractions, decimal fractions and percentages

6.6 Adding and subtracting fractions

We are learning to add and subtract like fractions.

Before we start

Amman and Isla have half a bar of chocolate each. Amman has more chocolate than Isla. Draw a diagram to show how this is possible.

We can use bar models to visualise fractions, making it easier to add and subtract.

Let's learn

One quarter means one *out of* four parts. We can write this as $\frac{1}{4}$.

one quarter

We call the digit below the line the *denominator* – this is the number of equal parts the whole bar has been split into.

$$\frac{1}{4}$$ numerator ← 1 denominator ← 4

We call the digit above the line the *numerator* – this is the number of parts that we are interested in (or the number of parts we are getting).

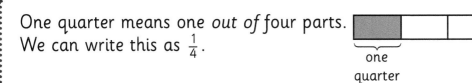

one quarter + two quarters = three quarters

$$\frac{1}{4} \quad + \quad \frac{2}{4} \quad = \quad \frac{3}{4}$$

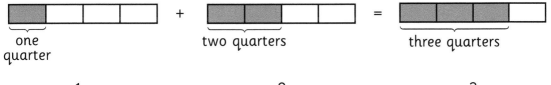

We can easily add like fractions – fractions with the same denominator.

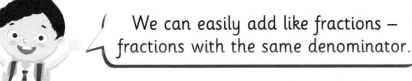

We can easily subtract like fractions also.

6

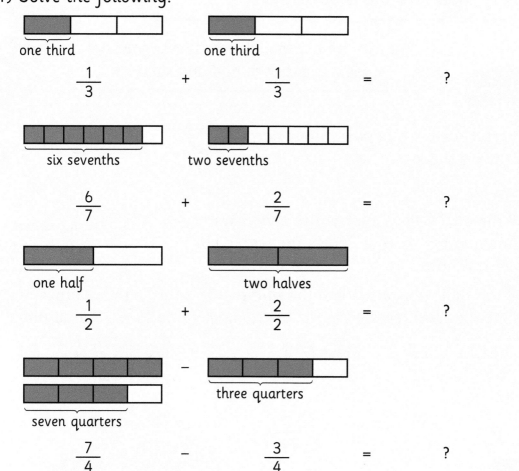

three quarters − two quarters = one quarter

$$\frac{3}{4} - \frac{2}{4} = \frac{1}{4}$$

Let's practise

1) Solve the following:

one third

$$\frac{1}{3}$$ + one third $$\frac{1}{3}$$ = ?

six sevenths

$$\frac{6}{7}$$ + two sevenths $$\frac{2}{7}$$ = ?

one half

$$\frac{1}{2}$$ + two halves $$\frac{2}{2}$$ = ?

seven quarters

$$\frac{7}{4}$$ − three quarters $$\frac{3}{4}$$ = ?

2) Draw bar models to help solve the following:

a) $\frac{2}{5} + \frac{1}{5} =$ ____

b) $\frac{7}{8} - \frac{3}{8} =$ ____

c) $\frac{5}{10} + \frac{5}{10} =$ ____

d) $1 - \frac{1}{3} =$ ____

e) $\frac{3}{4} + \frac{3}{4} =$ ____

3) Draw diagrams to solve the following problems:

a) Finlay eats three sixths of a pizza and Isla eats two sixths of a pizza. How much pizza have they eaten altogether?

b) Amman has eight tenths of a cake. He eats three tenths. How much does he have left?

c) Nuria cuts out and keeps five eighths of a piece of A4 paper. Finlay cuts out and keeps seven eighths of a piece of A4 paper. How much paper have they kept altogether?

d) Isla has one and three quarter bars of chocolate. She eats five quarters. How much chocolate does she have left?
Hint: Think about how to split a whole into quarters.

CHALLENGE!

Amman is trying to work out how much pizza he and Isla have altogether:

We can't add one quarter and one half because they are not like fractions.

$$\frac{1}{4} \quad + \quad \frac{1}{2} \quad = \quad ?$$

Help the children work out how much pizza they have altogether. What did you need to do to help them?

Work out the following:

Can we use what we know about equivalence to help us?

$\frac{1}{3} + \frac{1}{6} =$

$\frac{7}{8} - \frac{1}{4} =$

$\frac{1}{2} + \frac{4}{12} =$

$\frac{9}{10} - \frac{2}{5} =$

6.7 Calculating a fraction of a value

> We are learning to solve problems by calculating a fraction of a value.

Before we start

Draw a bar model to show how you could calculate:
- $\frac{2}{3}$ of 42
- $\frac{3}{4}$ of 96

> We can use bar models to visualise and solve fraction problems.

Let's learn

Three fifths of the crowd at a football match support Ayr. The rest of the crowd supports Dundee. There are **4750** *supporters altogether. How many Dundee supporters are there?*

fifths tells us how many parts to split the bar into:

three tells us how many parts relate to Ayr supporters.

total crowd

Ayr supporters   Dundee supporters

Therefore, two parts must relate to Dundee supporters:

4750 tells us the whole amount to be shared out.

4750 supporters ÷ 5 = 950 supporters:

4750
total crowd

$\frac{1}{5}$ of 4750 = 950	$\frac{1}{5}$ of 4750 = 950	$\frac{1}{5}$ of 4750 = 950	$\frac{1}{5}$ of 4750 = 950	$\frac{1}{5}$ of 4750 = 950

Ayr supporters   Dundee supporters

Two out of the five parts are Dundee fans so we only count the supporters in these parts.

2 × 950 supporters = **1900** Dundee supporters:

4750
total crowd

950	950	950	950	950

Ayr supporters   Dundee supporters
2850   1900

Let's practise

1) Use the bar models to work out the following:

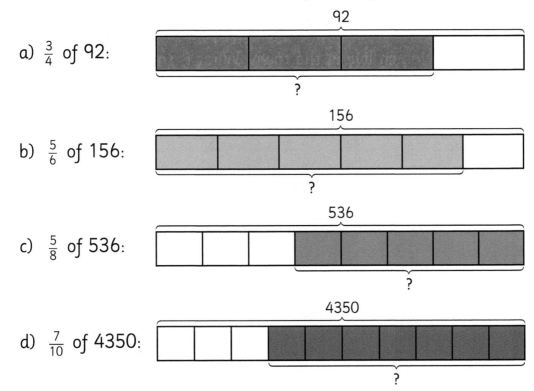

a) $\frac{3}{4}$ of 92:

92

?

b) $\frac{5}{6}$ of 156:

156

?

c) $\frac{5}{8}$ of 536:

536

?

d) $\frac{7}{10}$ of 4350:

4350

?

2) Make up a word problem for each of the bar models in question 1.

3) Draw a bar model to solve each of the following problems:
   a) Mark is taking part in a sponsored cycle from Glasgow to Paris.
   The total distance is 1120 km. He has travelled $\frac{1}{4}$ of the journey
   so far. How far has he still to cycle?
   b) A paint manufacturer produces green paint by mixing blue and
   yellow. $\frac{5}{9}$ of the mix is blue paint. How much yellow paint is required
   to make 4500 litres of green paint?
   c) Ava's Fitbit shows that she has taken 6885 steps by 6 pm. She did
   $\frac{2}{5}$ of the steps in the afternoon. How many steps did she take in
   the morning?

6

The children have been told that $\frac{4}{7}$ of the population of their local town are women. 480 women live in the town. What is the total population of their town? They draw a bar model to help:

?

Total population

Women
480

Men
?

We can't solve this because we don't know the whole amount in this problem!

a) Help the children work out the total population of their town.

b) Solve the following bar model problems:

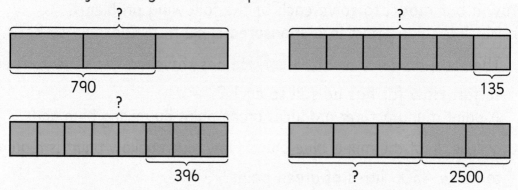

?

790

?

135

?

396

?

2500

c) Make up a word problem for each of the bar models.

6 Fractions, decimal fractions and percentages

6.8 Comparing numbers with two decimal places

We are learning to compare and order numbers with two decimal places.

Before we start

Isla has been asked to write a list of five numbers that are larger than 9·1 but smaller than 9·2 in order from smallest to largest. Help her write the list.

Decimal fractions help us record measurements with greater accuracy.

Let's learn

0·02 =   **two hundredths**   0·06 =   **six hundredths**

We can compare these two numbers by saying:

0·02 is smaller than 0·06 or ***0·06 is larger than 0·02***

Finlay has been asked to compare these two numbers: | 0·14 | | 0·08 |

0·14 has four hundredths and 0·08 has eight hundredths, so I can say

0·08 is larger than 0·14

I don't think that's right. Let's draw a bar for each number to check.

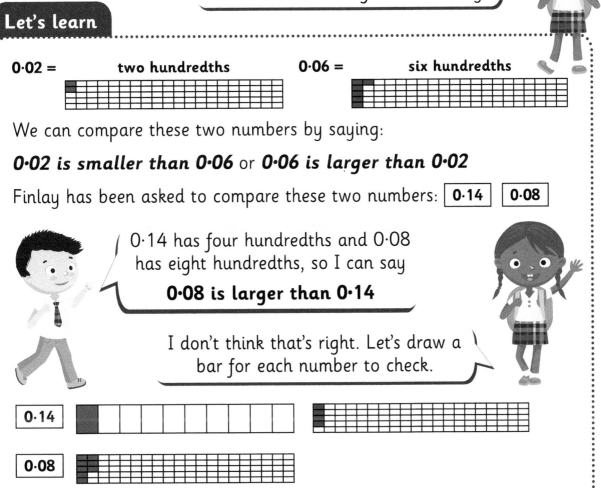

0·14

0·08

0·14 is actually one tenth of a bar plus four hundredths, so what Finn should have said is:

0·14 is larger than 0·08.

0·14 is the same as 14 hundredths, which is larger than eight hundredths.

Let's practise

1) Write a statement using decimal numbers to compare each of the bar models below.

a)

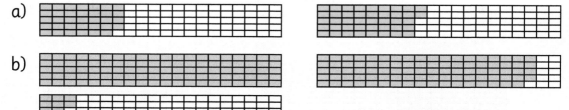

b)

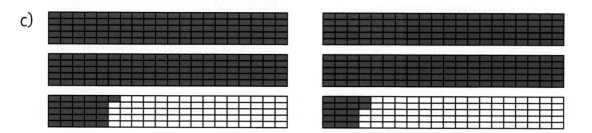

c)

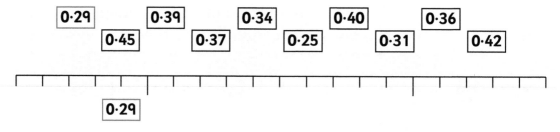

2) a) Copy the number line below and write the numbers in the correct positions. The first one has been done for you.

0·29		0·39		0·34		0·40		0·36

0·45		0·37		0·25		0·31		0·42

0·29

b) Make up four statements using the numbers above. For example,

0·29   is larger than   0·25

c) The athletes for the 100 m final ran the times below. List the athletes in the correct order from first to last.

GB 09:91 sec
Canada 10:05 sec
China 10:34 sec
USA 09:85 sec

Jamaica 09:79 sec
France 10:12 sec
South Africa 10:29 sec
Australia 10:43 sec

CHALLENGE!

There appears to be something wrong with the children's Fortday scoreboard. Can you reorder them correctly?

Name	Score	Position
Isla123	27·65	1st
1Amman1	28·34	2nd
Finlay999	28·07	3rd
Nuria247	27·70	4th
James007	28·19	5th
F.I.N.N.0·8	27·53	6th

Four more players join the game:

Mark101   Ted22   Bjorn2008   NewPlayer1

All four players score enough to put them into 2nd, 3rd, 4th and 5th place. What might each of them have scored?

What is the difference between the scores for first and last?

6 Fractions, decimal fractions and percentages

6.9 Percentage

We are learning to understand and use the term *percentage*.

Before we start

What fraction of these bars have been shaded?

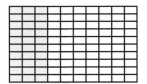 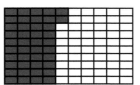 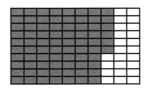

A percentage is another way of stating a fraction out of one hundred.

Let's learn

The word *percent* means *out of 100*. 'Per' means *out of* and 'cent' means *100*.

The percentage sign is made up of the digits that make 100: **%**

A percentage is another way of writing hundredths. Any fraction out of one hundred (hundredths) can be written as a percentage:

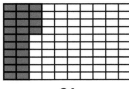

$$\frac{24}{100} \quad = \quad \textbf{24} \ out \ of \ \textbf{100} \quad = \quad \textbf{24\%}$$

twenty-four hundredths   twenty-four percent

$\frac{24}{100}$ means 24 *out of* 100 parts or 24%. They have exactly the same value.

Let's practise

1) The seating plan for six cinema screens shows seats that are occupied (red) and seats that are empty (green).

   a) Write the amount of occupied seats as both a fraction and a percentage:

   i) 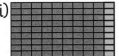   ii)    iii)    iv)

   b) What is the percentage of empty seats for each cinema?

2) A percentage can be written as a fraction out of 100. Convert the following percentages into fractions (the first one has been done for you):

   a) 1% = 1 out of 100 = $\frac{1}{100}$

   b) 13% = ___ out of 100 = ___

   c) 72% = ___ out of 100 = ___

   d) 85% = ___ out of 100 = ___

CHALLENGE!

Nuria believes that any fraction can be converted into a percentage, not just hundredths:

One half can be converted into 50 hundredths, so we can say that:
$\frac{1}{2}$ equals 50%

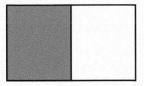   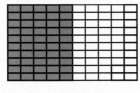

$\frac{1}{2}$ = $\frac{50}{100}$ = 50 *out of* 100 = 50%

one half   50 hundredths   50 percent

Convert the following fractions into percentages using Nuria's stragey:

$\frac{1}{4}$ $\frac{1}{5}$ $\frac{1}{10}$ $\frac{1}{25}$

Can you find other fractions that can be converted into a percentage?

6 Fractions, decimal fractions and percentages

6.10 Converting fractions to percentages

We are learning to convert fractions into percentages.

Before we start

Convert the following fractions into hundredths: $\frac{2}{10}$ $\frac{3}{5}$ $\frac{1}{4}$

Equivalence can help us to convert a fraction into a percentage and vice versa.

Let's learn

If a percentage is the same as a fraction out of 100, it should be easy to convert any fraction to a percentage.

Only if the fraction can be easily converted into hundredths!

We can use equivalence to convert a fraction to hundredths.

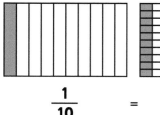

$$\frac{1}{10} \quad = \quad \frac{10}{100} \quad = \quad 10 \; out \; of \; 100 \quad = \quad 10\%$$

one tenth   ten hundredths   ten percent

One tenth is equivalent to ten hundredths. Ten hundredths means 10 out of every 100. So $\frac{1}{10}$ is equal to 10%.

We can also convert a percentage into a fraction:

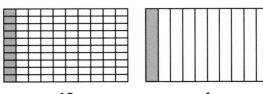

10% = **10** *out of* **100%** = $\frac{10}{100}$ = $\frac{1}{10}$

ten percent   10 hundredths   one tenth

Let's practise

1) Convert the following fractions to percentages:

a)

$\frac{1}{4}$ =   _____   = _____ %

b)

$\frac{3}{5}$ =   _____   = _____ %

c)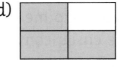

$\frac{9}{10}$ =   _____   = _____ %

d)

$\frac{3}{4}$ =   _____   = _____ %

2) Finlay has been asked to rewrite the following signs using fractions in their simplest form. Can you help him?

a) **SALE** 70% off   = $\frac{?}{?}$ = $\frac{?}{?}$ =   **SALE** _____ off

b) **SALE** 40% off   = $\frac{?}{?}$ = $\frac{?}{?}$ =   **SALE** _____ off

c) **SALE** 25% off   = $\frac{?}{?}$ = $\frac{?}{?}$ =   **SALE** _____ off

d) **SALE** 75% off $= \dfrac{?}{?} = \dfrac{?}{?} =$ **SALE** ____ off

CHALLENGE!

Isla explains to Amman that some fractions are more difficult to convert into a percentage. She gives him the following fraction cards to investigate:

$\dfrac{3}{5}$ $\dfrac{1}{3}$ $\dfrac{7}{10}$ $\dfrac{5}{8}$ $\dfrac{3}{6}$ $\dfrac{4}{20}$ $\dfrac{1}{7}$ $\dfrac{12}{50}$ $\dfrac{6}{8}$ $\dfrac{1}{100}$

Some fractions convert into a percentage with a decimal point like 33·33%.

Can you help me find the fractions that can't be easily converted to a percentage?

a) Convert the fractions above into percentages.

b) Explain why some fractions are more difficult to convert.

6 Fractions, decimal fractions and percentages

6.11 Percentage calculation

We are learning to solve problems by calculating the percentage of a value.

Before we start

A packet contains 20 sweets. Four tenths of them are orange-flavoured. How many orange-flavoured sweets are there in the bag?

A percentage of an amount can be calculated in the same way as a fraction of an amount.

Let's learn

There are 30 children in our class. 30% of the class are boys. How many boys are there?

We can use a bar model to visualise the problem:

whole class
100%
30 children

10%	10%	10%	10%	10%	10%	10%	10%	10%	10%

boys
30%

girls
70%

The whole bar represents the whole class: 30 children.

We can split the bar into 10 parts of 10% each.

To work out 30% of the children we simply divide the 30 children equally between the 10 parts and then count three of these parts:

whole class
30 children

3	3	3	3	3	3	3	3	3	3

boys
9 boys

girls
21 girls

There must be nine boys in the class.

6

1) The bar models below show the percentage of girls and boys that attend various after school clubs.

Calculate how many girls and boys attend each club:

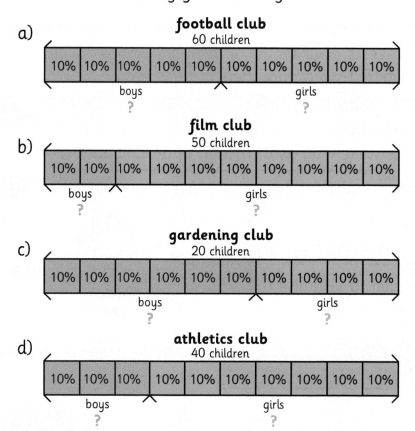

a) **football club**
60 children

| 10% | 10% | 10% | 10% | 10% | 10% | 10% | 10% | 10% | 10% |

boys ?   girls ?

b) **film club**
50 children

| 10% | 10% | 10% | 10% | 10% | 10% | 10% | 10% | 10% | 10% |

boys ?   girls ?

c) **gardening club**
20 children

| 10% | 10% | 10% | 10% | 10% | 10% | 10% | 10% | 10% | 10% |

boys ?   girls ?

d) **athletics club**
40 children

| 10% | 10% | 10% | 10% | 10% | 10% | 10% | 10% | 10% | 10% |

boys ?   girls ?

2) Draw a bar model to solve each of the following problems:
   a) Finlay has £70 saved in his piggybank. He spends 10% of it on a new book. How much was the book?
   b) A farmer has 140 animals on his farm. 40% of them are sheep. How many sheep does he have on the farm?
   c) A pair of trainers usually cost £50. Isla buys them in a sale with 20% off. How much does she save?
   d) 80% of the children in a school attend the summer fayre. How many children attended the fayre if there are 300 children in the school altogether?

CHALLENGE!

a) Amman has been asked to make up two percentage problems for the following bar model:

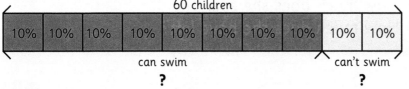

| 10% | 10% | 10% | 10% | 10% | 10% | 10% | 10% | 10% | 10% |

> 80% of the children in a swimming club can swim the backstroke. There are 60 children in the club altogether. How many children can swim the backstroke?

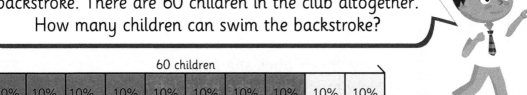

60 children

| 10% | 10% | 10% | 10% | 10% | 10% | 10% | 10% | 10% | 10% |

can swim   can't swim
?   **?**

Write another question for the bar model using a different whole amount.

Solve both the problems.

b) Make up and solve two questions for the following bar model:

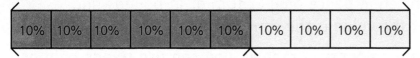

| 10% | 10% | 10% | 10% | 10% | 10% | 10% | 10% | 10% | 10% |

7 Money

7.1 Money problems using the four operations

We are learning to solve problems using £ and p notation.

Before we start

Nuala has a £10 note, a £5 note, one £2 coin and five 20p coins.

She spends £12·50 on a new bag and £1·20 on a purse.

a) How much does she spend?
b) How much does she have left?

We are using our knowledge of £, p and decimal notation.

Let's learn

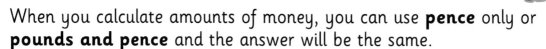

Money can be written in pence only (236p) or in pounds and pence (£2·36) but the value is the same.

When you calculate amounts of money, you can use **pence** only or **pounds and pence** and the answer will be the same.

Laura buys four pens that cost £2·95 each, how much does she spend **altogether**?

£2·95 × 4 = £11·80

295p × 4 = 1180p

Let's practise

1) Marie buys a jacket that costs £52·25.

   She uses three £20 notes to pay.
   a) Using £ and p, how much change does Marie get?
   b) List one possible combination of notes and coins that she might receive in her change.

2) Lewis wants to buy three tickets for a film that each cost £6·75.
   He has two £5 notes, two £1 coins and seven 50p coins.
   a) How much money does Lewis have?
   b) Does he have enough to buy the tickets?

3) Miss Noon wants to take her Primary 5 class of 26 pupils to Deep Sea Aquarium.
   An adult ticket is £5·50 and a child ticket is £3·30.
   a) How much will it cost for three adults and 26 pupils?
   b) She has six £20 notes, how much change will she receive?

   c) List the possible notes and coins she might receive in her change.

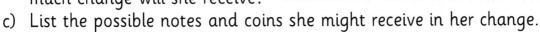

CHALLENGE!

Lauren, Paul and Matthew each receive £5 per week for completing their chores.
At the end of four weeks they have saved some money and spent some of it. They count their money and altogether they have eight £5 notes, four £2 coins, two £1 coins and four 20p coins.

a) How much do they get, in total, for completing their chores?
b) How much do they actually have?
c) If Lauren spent £2·50 and Paul spent £2·70, how much did Matthew spend?

7 Money

7.2 Budgeting

We are learning to compare costs and keep to a budget.

Before we start

Cosmo wants to buy a new laptop that costs £445.
He saves £30 per week.

a) How many weeks will he need to save for before he can afford the laptop?

b) He notices that the store Compusave are offering £50 if you trade in your old laptop.

How many weeks sooner will he be able to buy his laptop if he does this?

When we compare costs, we can save money and keep to our budget.

Let's learn

Most people need to **budget** their money in order to decide what they can afford to buy.

Different retailers often sell the **same** items, but they charge **different** prices for them.

Retailers can also have offers on certain items that can make the product cheaper if you buy more than one.

Therefore, you can save money if you **compare the costs** of the items you want to buy.

Finlay has £10 and wants to buy football stickers for his album.

He looks in two different shops …

Wentworths sell individual packets for £1·50 and if you buy two packets you can get the third free.

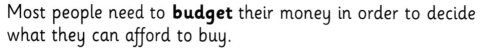

(1 = £1·50, 2 = £3·00 and 3 = £3·00)

Convenience Corner sell each packet for £1·30 but currently offer 'buy one get one half price'.

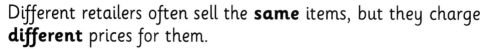

(1 = £1·30, 2 = (£1·30 + 65p = £1·95)

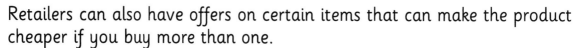

At which can he get the best value?

Answer:

At Wentworths he can get nine packs for £9. At Convenience Corner he can get 10 packs for £9·75. He can buy the most packs at Convenience Corner and get the best value.

Let's practise

1) Lauren wants to buy a laptop.

If she goes into IT Supplies, the laptop is £450 and she will get a £30 student discount.
If she buys the same laptop online for £450 she will get a £25 student discount but will have to pay £4·75 for delivery.

a) Where should she buy her laptop?   b) Will she save any money?

2) Paul is learning to drive.

Road Runners offer lessons for £28 per lesson or a block of 10 lessons for £260.

Diamond Drivers offer lessons for £30 each but every fifth lesson is free.
U Can Drive offer lessons for £25 each.

a) If Paul books 10 lessons, which company offers the best deal?
b) What is the **least** he will pay for 10 lessons?
c) What is the **least** he will pay per lesson?

CHALLENGE!

Nuria loves ice skating and wants to attend The Ice Palace for lessons.

Skate hire is £3 per hour. Each lesson costs £5·50 for one hour.

A block of six lessons costs £30 with £2·50 off skate hire.

If you pay monthly it costs £24 and includes free skate hire.

a) What is the **least** Nuria will pay for each lesson including her skate hire?

b) What is the **most** she could pay for each lesson including her skate hire?

c) If Nuria's mum was booking a block of six lessons, what is the **total** she would budget for including her skate hire?

7.3 Profit and loss

We understand and can use the terms profit and loss.

Before we start

Charlotte bought four chairs at the market for £15 each.

If she sells them for a total of £80, how much profit does she make per chair?

Profit and loss are terms to describe how much money you make or lose when you sell something.

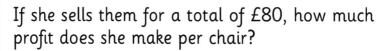

Let's learn

Profit = selling price – buying price (where selling price is bigger)

Loss = buying price – selling price (where buying price is bigger)

A retailer bought eight bunches of grapes for £6·20 and sold them for £1·20 per bunch.

How much did he make?

Total selling price = £1·20 × 8

= £9·60

Profit = £9·60 – £6·20

= £3·40

If the retailer sold each bunch for 70p per bunch, would he make a profit or a loss?

Total selling price = 70p × 8

= £5·60

Loss = £6·20 – £5·60

= £0·60 or 60p

Let's practise

1) Mr Green the grocer bought 30 bunches of bananas for £18·00.
   He sold 22 bunches for 80p each. He didn't sell the remaining bunches and had to give them away.
   a) Did Mr Green make a profit or a loss?
   b) How much did he make or lose?

2) Primary 6 wanted to raise some money for their class trip, so they decided to make cakes.

   If the ingredients cost them £4 altogether and they made 24 cakes, how much would they have to sell the cakes for if they want to make £12 profit?

3) Jo Joiner made a treehouse for Mr Hunter's children.
   It cost him £50 to buy the wood, £15 to buy all the other materials and he charged £25 for his time.
   a) If he charged Mr Hunter £80 did he make a profit or a loss and by how much?
   b) If he wanted to make a profit of £25, how much would he have to charge Mr Hunter for the treehouse?

CHALLENGE!

Isla and Amman have set up a lemonade stall to raise some money to buy new sports equipment for their school.

The ingredients for eight cups are listed below:

one bottle of soda water   £0·60
six lemons   £3·00
125 g granulated sugar   £1·60 per 500 g

a) How much would each cup of lemonade cost?
b) How much will they need to sell each cup for to make a 10p profit per cup?
c) If they sell 20 cups and make £6 profit, how much will they need to sell each cup for?

d) If Amman wants to make a £20 profit, how many cups of lemonade must they sell and how much should they charge per cup?

7 Money

7.4 Discounts

We are learning about discounts.

Before we start

Finlay's mum gave him £15 to buy stationery for school.

He needs a pencil case, two pencils, two pens, an eraser, a ruler and a sharpener.

He also wants a set of gel pens and some coloured pencils.

Can he get everything he needs *and* wants for £15?

Pencils 75p each, Pens £1.20 each.

Pencil case with 2 pencils, 2 pens, 2 erasers, 1 sharpener, 1 ruler and coloured pens £14.99.
Pack of gel pens £6.50.
Ruler £2.50.
Pencil case £4.50.

A discount is when money is taken off the original value of an item.

Let's learn

Retailers offer **discounts** on items using **fractions** and **percentages** of the original cost to calculate the new, **lower price**.

A games console costing £180 has 50% off, how much will it cost?

How much will it cost in the sale?

Original price   £180

Discount (50%) = $\frac{1}{2}$ of £180 = £90

New price   = £180 – £90 = **£90**

Let's practise

1) Find the cost of the following items after the discount:

a) Original cost   £28
   Discount   50% off

b)    Original cost   £90

              Discount $\frac{1}{2}$ price

2) Nuala joins a ballet class that costs £120 per term.

   She is given an introductory offer of 10% off.
   a) What is the value of Nuala's discount?
   b) How much does she pay for the ballet class?

3) Mairi-Louise bought a new pair of shoes in the sale.

   She paid £75, which included a 50% discount.
   How much did the pair of shoes cost originally?

CHALLENGE!

Kenny bought two new shirts. Each had a 10% discount. He saved a total of £5.

a) What was the original cost of each shirt?

b) How much did he pay in total?

7 Money

7.5 Credit, debit and debt

We are learning about the terms credit, debit and debt.

Before we start

Match the words to their correct definitions.

Debt	**Money taken out of a bank account**
Credit	**A sum of money that is owed**
Debit	**A sum of money that is received or money that is given in advance**

Debit and credit cards can be used to purchase items.

Let's learn

Many people use either a debit or credit card when paying for their purchases.

Banks give **debit cards** to customers when they open a bank account and they can be used to pay for goods and services.

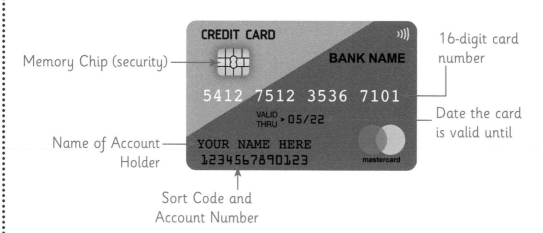

Memory Chip (security)

CREDIT CARD

BANK NAME

5412 7512 3536 7101

VALID THRU ▸ 05/22

16-digit card number

Date the card is valid until

Name of Account Holder

YOUR NAME HERE
1234567890123

mastercard

Sort Code and Account Number

When you use your debit card you need a **personal identification number** (PIN) which is known only by you and is a four-digit number.

A **credit card** is used in the same way as a debit card when you buy something. When you use a credit card you are borrowing money and will be charged interest on the amount you owe. A debit card takes money directly out of your bank account.

Let's practise

1) Discuss the following:
   a) Why should the **PIN** be known by the card holder only?
   b) Is using a card better than using money? Why?
   c) Why does each person need an **account number**?

2) What is the difference between a **debit card** and a **credit card**?

3) Write three **advantages** and three **disadvantages** of using a **credit card**.

   Discuss these with your partner.

CHALLENGE!

Work with a partner to investigate the following banking terms:
- ATM machines
- Overdrafts
- Direct debit

8 Time

8.1 Reading and writing 12-h and 24-h time

We are learning to read and write 12-hour and 24-hour time.

Before we start

Write the following times in words:

a)

b)

The 24-hour clock is one way of recording the time from midnight to midnight.

Let's learn

If the time is between 1·00 am and 12·59 pm, then 12-h and 24-h time are the same.

If the time is between 1·00 pm and 11·59 pm, then we add 12 hours to the time to show 24-h time.

Look at the time on this digital clock:

It is in the evening, but what hour is it?

The clock shows 19 hours and 30 minutes.

To change this time from 24 h to 12 h, take 12 from the hour: $19 - 12 = \textbf{7}$

We now know that the hour is 7, therefore the time is **1930** or **7·30 pm**.

Let's practise

1) Copy and complete the following table:

12-hour clock	24-hour clock
12·00 pm	
	0330
5·21 pm	
	0814
12·00 am	

2) Write the following times in both 12-h and 24-h notation:

a) Fourteen minutes past three in the afternoon.
b) Five minutes to four in the morning.
c) Twenty-five minutes to eleven at night.
d) Fifteen minutes to eight in the morning.
e) Midnight.
f) Three minutes to two in the afternoon.

3) Put the following times into the correct order:

a) 12·25 pm   b) 0200   c) 8 pm
d) 2134   e) Ten in the morning

CHALLENGE!

On a digital clock displaying 24-h time, how many times does the number 5 appear over the whole day?

22:45

8 Time

8.2 Converting units of time using fractions

We are learning to convert units of time using fractions.

Before we start

Work with a partner to match the following times:

Seven o'clock

Quarter past four

Half-past six

To convert something, you change it into a different form.

Let's learn

To convert units of time you need to know the following facts:

60 seconds = one minute

60 minutes = one hour

15 minutes = $\frac{1}{4}$ of an hour

30 minutes = $\frac{1}{2}$ an hour

45 minutes = $\frac{3}{4}$ of an hour

If the maths lesson started at 11·55 am and lasted for $\frac{3}{4}$ of an hour, what time did it finish?

*If the lesson began at 11·55 am and lasted for a total of **45 minutes**, then it would finish at 12·40 pm.*

Let's practise

1) Nuria's bus leaves at 8·15 am. The journey lasts for 15 minutes. Does Nuria arrive at school at:

   a) 0845   b) $\frac{1}{2}$ past eight   c) Twenty minutes past eight

2) Finlay is watching a film that starts at 1530 and finishes at 1645. How many minutes does it last:

   a) 60 minutes   b) 95 minutes   c) 75 minutes

3) Mr Smith's watch is showing 7·15 am. If his watch is 30 minutes slow, what should the time be?

   a) 7·30 am   b) 7·00 am   c) 7·45 am

CHALLENGE!

Amman and Isla went for a walk in the country.

They set off at 1530 and walked for $\frac{1}{4}$ of an hour before stopping for a drink.

They began walking again at 1600 and walked

for another $\frac{1}{2}$ an hour.

How many minutes were they walking for in total?

8 Time

8.3 Calculating time intervals using timetables

We are learning to calculate time intervals using timetables.

Before we start

The timetable shows the departure time of each bus.

🚌 Bus Timetable
Look for details on each station, including the address and local transit connections, below on the page.

Mondays to Saturdays except public holidays										Sundays public holidays				
0754	0845	0945	1045	1145	1245	1345	1445	1545	1645	1055	1355	1455	1545	1655
0750	0850	0950	1050	1150	1250	1345	1450	1550	1650	1055	1355	1255	1845	1855
0755	0855	0955	1055	1155	1255	1345	1455	1555	1655	1050	1150	1250	1350	1450
0854	0945	1045	1145	1245	1345	1445	1545	1645	1745	1345	1445	1545	1645	1745
0850	0950	1050	1150	1250	1350	1450	1550	1650	1750	1350	1450	1550	1650	1750
0911	1011	1111	1211	1311	1411	1511	1611	1711	1811	1411	1511	1611	1711	1811
▼	0850	▼	1050	▼	1255	▼	1450	▼	1650	1345	1445	1545	1645	1745
0750	▼	0950	▼	1150	▼	1345	▼	1550		▼	▼	▼		
▼	0855	▼	1055	▼	1255	▼	1455	▼	1655	1611	1711	1811		
0755	▼	0955	▼	1155	▼	1345	▼	1555		1455	1555	1655		
0750	0850	0950	1050	1150	1250	1345	1450	1550		1455	1555	1655		

1) If I arrive at the station at 1200, what is the next bus I can board?

2) What time does the last bus leave from the station?

3) Are there any days that the bus times might be different?

An interval is a distinct measure of time.

Let's learn

To calculate an interval of time, you need to know a start time and an end time.

If a film **starts** at 1620 and **finishes** at 1750, how long does it last?

1620 1630 1640 1650 1700 1710 1720 1730 1740 **1750**

So, if the film starts at 1620 and finishes at 1750 it lasts for

90 minutes in total, which is the same as $1\frac{1}{2}$ **hours**.

Let's practise

1) Use the timetable to answer the following questions:

a) Paul is getting the No. 47 into town.
If he catches the 1520 and gets off the bus at the last stop, how long is his journey?

b) Mrs Rochdale always gets the No. 49 to the market.
If she catches the 1615 and her journey lasts eight minutes, what time will she arrive?

2) Finlay and his mum get the train to Aberdeen that leaves Edinburgh at 1025 and arrives in Aberdeen two hours and 10 minutes later.

What time did Finlay and his mum arrive in Aberdeen?

CHALLENGE!

A new TV channel, JTV is being set up.

They have to plan the TV programmes from 6 pm to 12 midnight.

There are a number of guidelines that need to be followed:

- every $\frac{1}{2}$ hour there needs to be an advert break

- 5% of the time should contain news and weather

- a feature film of $1\frac{1}{2}$ hours must be shown

- advert breaks last two minutes

    a) What is the total time JTV is on air?

    b) How many minutes should contain news and weather?

    c) How long does the film last, including the adverts?

8 Time

8.4 Measuring time

We are learning to use different instruments of measure to calculate time.

Before we start

List five things that would take you one minute to do.
Discuss your answers with a partner.
List as many instruments you would use to measure time as you can.
Discuss your answers with a partner.

Time can be measured in a variety of different ways.

Let's learn

Some of the instruments we may use to measure time are: clocks, stopwatch, hourglass, calendar and sundial. The instruments that we use to measure time will depend on the total time of the activity or event.

The unit of time we use to measure time will also depend on the activity or event. Units of time include: seconds, minutes, hours, days, weeks, years…

Let's practise

1) Using seconds, minutes, hours or days, estimate how long the following would take:

a) Brush your teeth b) Fly to Australia
c) Blow your nose d) Run a marathon
e) Read a book

2) Which of the following would you use to time a race?

   a) Calendar   b) Sundial   c) Stopwatch

3) Which of the following would not be used to measure time?

   a) Hourglass   b) Watch   c) Candle
   d) Diary   e) Ruler

4) Create a list of timing devices and the units of time they measure. For example, a standard clock measures hours, minutes and seconds. How many more can you list?

CHALLENGE!

Copy and complete the following table:

Activity	Estimated time	Actual time	Unit of time	Instrument of measure
Write my name				
Sharpen five pencils				
Hop 50 times				
Walk round the playground twice				
Add more of your own activities				

8 Time

8.5 Speed, distance and time calculations

We are learning to calculate speed, distance and time.

Before we start

Using the formula **distance = speed × time** calculate the following:
If Mark was travelling at 40 mph for four hours, what distance did he travel in total?

Speed is measured as **distance** travelled per unit of **time**.

Let's learn

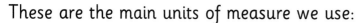

These are the main units of measure we use:

Distance – metres, miles or kilometres.

Speed – metres, miles or kilometres per minute or hour.

Time – seconds, minutes or hours

To calculate distance, we can use the formula Distance (D) = Speed × Time

For example:

Finlay rides his bike at **40 mph** for $3\frac{1}{2}$ **hours**, how far does he cycle?

This is a two-part calculation:

First, calculate complete hours: D = 40 × 3

D = 120 miles

Then calculate the fraction of an hour – if he travels 40 mph then he will travel half that in 30 minutes, which is **20 miles**.

D = 120 miles + 20 miles

D = 140 miles

Let's practise

1) What unit of measure would you use to estimate the distance travelled by:
   a) a snail slithering for two minutes
   b) a car driving at 45 mph
   c) a girl hopping for five minutes

2) What unit of time would you use to estimate the duration of:
   a) jumping in the air twice
   b) flying to Spain
   c) running across the playground

3) Matthew was driving for $2\frac{1}{2}$ hours at 50 mph.

   What distance did he travel in total?

4) Charlotte cycled 25 mph for $3\frac{1}{2}$ hours.

   How far did she cycle?

CHALLENGE!

Mr Jones drove to work this morning.
He drove at 45 mph.
He left the house at 7·30 am and arrived at his office at 9 am.
How far away is his office?

8 Time

8.6 Time problems

We are learning to solve time problems using the four operations.

Before we start

Answer true or false for the following:
a) There are 90 minutes in an hour.
b) 0230 is the same as 2·30 am.
c) There are 12 hours in a day.
d) The minute hand is longer than the hour hand.

You can use +, −, × and ÷ to solve time problems.

Let's learn

When adding, subtracting, multiplying and dividing units of time, you need to remember that there are 60 seconds (secs) in one minute and 60 minutes (mins) in one hour.

If Isla swam four lengths of the pool and each length took her two minutes and 35 seconds, how long did it take her to swim four lengths altogether?

2 mins 35 secs + 2 mins 35 secs + 2 mins 35 secs + 2 mins 35 secs
= **10 mins 20 secs**

2 mins 35 secs × 4 = **10 mins 20 secs**

You can calculate this answer either by adding or multiplying.

Let's practise

1) There were four members of the relay team and their times were as follows:

John – 3 mins 15 secs   Jim – 2 mins 20 secs
Jason – 3 mins 5 secs   Jett – 2 mins 40 secs

How long did their race last in total?

2) It takes Simon 25 minutes every morning to walk to work and 25 minutes to walk back.

If he works five days a week, how many minutes does he spend walking every week?

3) Tom, Tina, Tony and Tess are all sharing the games console.

They have two hours to split between the four of them.

a) How long will each of them have to play a game?
b) If their friend Antonia joins them, how long would they each get now?
c) If Attila also joins in, how long would they each get?

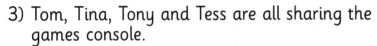

CHALLENGE!

If there are 30 days in April, can you work out the following facts:

a) How many hours are there?
b) How many minutes are there?
c) How many seconds are there?

9 Measurement

9.1 Estimating and measuring length

We are learning to estimate and measure lengths to one decimal place.

Before we start

Use a ruler to measure each of these lines:

We can use decimal numbers to give measurements with greater accuracy.

Let's learn

We use different units of length to measure different lengths:

1 km = 1000 m   1 m = 100 cm   1 cm = 10 mm

100 m = $\frac{1}{10}$ km = 0·1 km   10 cm = $\frac{1}{10}$ m = 0·1 m

1 mm = $\frac{1}{10}$ cm = 0·1 cm

The distance from Ayr to Troon is between 8 and 9 kilometres:

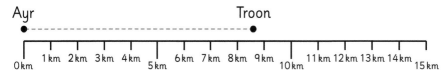

We can be more accurate if we zoom in on the line:

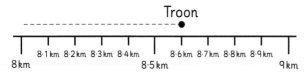

We can now say, more accurately, that the distance from Ayr to Troon is exactly 8·6 km or 8600 m.

This pencil is between 6 and 7 centimetres:

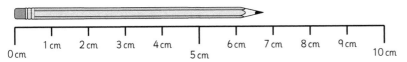

We can be more accurate if we zoom in on the ruler:

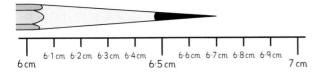

We can now say, more accurately, that the length of the pencil is exactly 6·7cm or 67mm.

Using decimals doesn't change the length we are measuring, it allows us to be more accurate.

Let's practise

1) Use a ruler to measure the length of each of the lines accurately. State the measurements in millimetres and centimetres (the first one has been done for you):

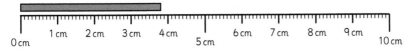

Length = 38mm or 3·8cm

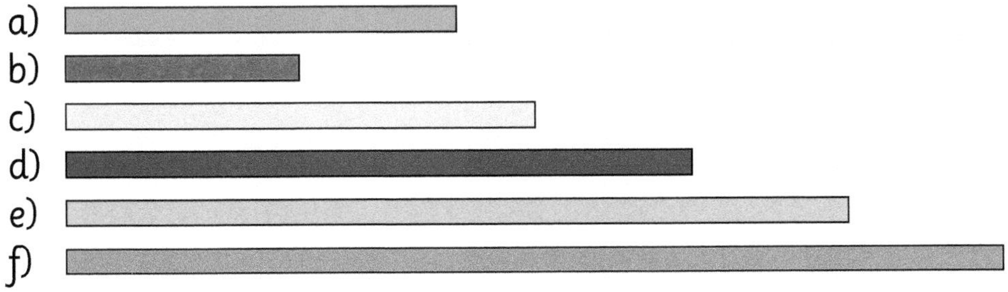

a)
b)
c)
d)
e)
f)

2) In a race, the distance each cyclist has travelled so far is shown in metres. Write each of the distances in kilometres (the first one has been done for you):

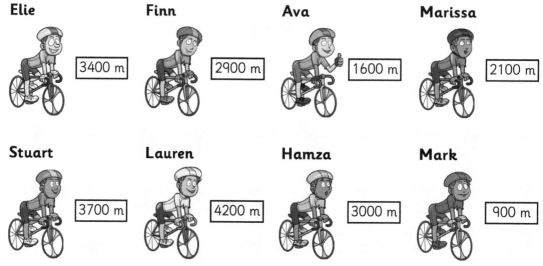

Elie   Finn   Ava   Marissa

3400 m   2900 m   1600 m   2100 m

Stuart   Lauren   Hamza   Mark

3700 m   4200 m   3000 m   900 m

Elie: 3400 m = 3·4 km

3) Amman uses a tape measure to measure the following objects in centimetres. Help Amman convert each measurement into metres (the first one has been done for you):

a) **Length of table = 180 cm = 1·8 m**
b) Length of smartboard = 240 cm
c) Length of window = 390 cm
d) Length of classroom = 670 cm
e) Height of door = 200 cm
f) Height of chair = 80 cm

☆ **CHALLENGE!** ....

a) Choose six items from your environment around you. Estimate their length in centimetres to one decimal place, then measure them accurately with a ruler or tape measure. Create a table to display your results.

b) Use chalk or a length of string. Draw lines (or cut pieces of string) that you estimate to be:

   • 1·5 m   • 2·4 m   • 3·2 m   • 0·8 m

Use a metre stick or tape measure to measure the lengths accurately and check how close you were.

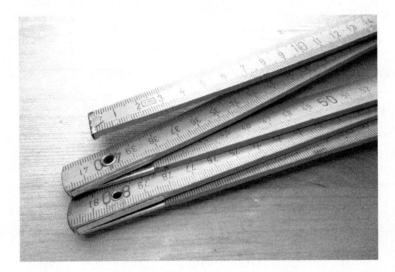

9 Measurement

9.2 Estimating and measuring mass

We are learning to estimate and measure the mass of an object to one decimal place.

Before we start

Find and measure the weight of items you estimate to have a mass of:

- 300 g
- 500 g

Calculate the difference between the estimate and actual mass for each item.

We can measure mass more accurately using decimal numbers.

Let's learn

1 kg = 1000 g

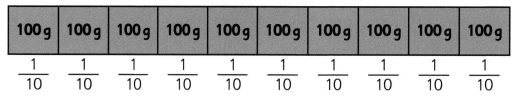

100 g	100 g	100 g	100 g	100 g	100 g	100 g	100 g	100 g	100 g
$\frac{1}{10}$	$\frac{1}{10}$	$\frac{1}{10}$	$\frac{1}{10}$	$\frac{1}{10}$	$\frac{1}{10}$	$\frac{1}{10}$	$\frac{1}{10}$	$\frac{1}{10}$	$\frac{1}{10}$

1 kilogram can be split into ten equal parts of 100 grams. 100 grams is one tenth of a kilogram. We can write this as a decimal:

$100\,g = \frac{1}{10}\,kg = 0{\cdot}1\,kg$

The weight of this box is between 2 and 3 kilograms:

We can be more accurate if we zoom in on the scale:

We can now say, more accurately, that the weight of the box is exactly 2.7 kg or 2700 g.

Using decimals doesn't change the weight of the object; it allows us to be more accurate when we write the weight.

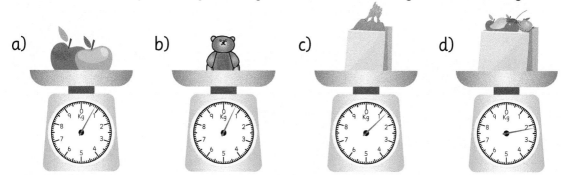

1) Write the mass of each of the objects below in both grams and kilograms:

a)

b)

c)

d)

2) a) Find two items in the environment around you that you estimate to have a mass of between 0·1 kg and 0·5 kg.

b) Now find two items that you think will have a mass between 0·6 kg and 1·0 kg.

c) Finally, find two items that you guess will have a mass of between 1·1 kg and 1·5 kg.

Create a table to record your estimates and then measure the items accurately using suitable scales.

Item	Estimated mass	Actual mass

CHALLENGE!

You will need:

- Suitable scales (hanging scales may work best)
- A selection of items to be weighed
- A bag

Can you fill a bag with items that total the following weights?

a) 0·6 kg   b) 0·9 kg   c) 1·2 kg   d) 1·5 kg   e) 2·3 kg

9.3 Estimating and measuring capacity

We are learning to measure capacity correct to one decimal place.

Before we start

How much water is contained in each of these measuring jugs?

What is the total volume of the two jugs?

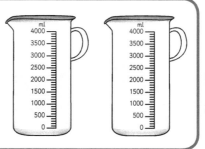

We can measure capacity more accurately using decimal numbers.

Let's learn

1 litre = 1000 ml

100 ml	100 ml	100 ml	100 ml	100 ml	100 ml	100 ml	100 ml	100 ml	100 ml
$\frac{1}{10}$	$\frac{1}{10}$	$\frac{1}{10}$	$\frac{1}{10}$	$\frac{1}{10}$	$\frac{1}{10}$	$\frac{1}{10}$	$\frac{1}{10}$	$\frac{1}{10}$	$\frac{1}{10}$

One litre can be split into 10 equal parts of 100 ml. 100 ml is one tenth of a litre. We can write this as a decimal:

100 ml = $\frac{1}{10}$ L = 0·1 L

This measuring jug contains between three and four litres of water:

We can be more accurate if we zoom in on the scale:

We can now say, more accurately, that the measuring jug is filled to exactly 3·6 L or 3600 ml.

Using decimals doesn't change the amount of water in the jug; it allows us to be more accurate when we write the volume of liquid.

Let's practise

1) How much liquid is contained in each of the measuring jugs below? Write the measurements in both millilitres and litres (the first one has been done for you):

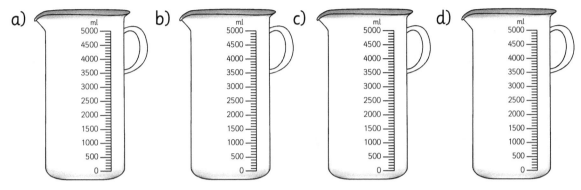

a)   b)   c)   d)

Millilitres: 800 ml
Litres: 0·8 L

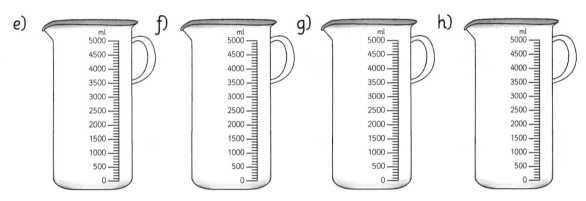

e)   f)   g)   h)

2) Use a measuring jug to measure out the following volumes of water, then write each measurement in litres (the first one has been done for you):

a) 100 ml = 0·1 L   b) 400 ml = _____
c) 600 ml = _____   d) 800 ml = _____
e) 1200 ml = _____   f)   1700 ml = _____

You may need to fill more than one measuring jug for some of these volumes of water!

CHALLENGE! ....

You will need:

- One full two-litre bottle of water
- One empty two-litre bottle of water
- funnel
- measuring jug
- to work with a partner

Repeat the instructions below for each of the following measurements:

a) 0·2 L   b) 0·5 L   c) 0·9 L

d) 1·3 L   e) 1·6 L   f) 1·8 L

i) Using estimation, pour water into the empty bottle. (Use the funnel while your partner holds the bottle steady).

ii) Empty the water into the measuring jug.

iii) Read the measurement.

This measures 0·6 L or 600 ml.

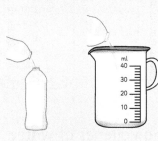  

iv) Copy and complete the table:

Container	Estimate	Actual	Difference
a			
b			
c			
d			
e			
f			

9 Measurement

9.4 Converting metric units

 Before we start

What is the volume of these cuboids?

We are learning to convert between metric units of volume and capacity.

We can convert cubic centimetres to millilitres and vice versa.

Let's learn

Volume is the space taken up by a 3D object.

The **volume** of this cube is one cubic centimetre (1 cm³):

1 cm, 1 cm, 1 cm

Capacity measures the amount of liquid a container can hold.

This container has a **capacity** of one millilitre (1 ml) of liquid:

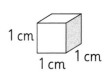
1 cm, 1 cm, 1 cm

We can see that the space taken up by one cubic centimetre is equal to the space taken up by one millilitre of liquid, so:

$$1 \text{ cm}^3 = 1 \text{ ml}$$

This cube is made up of 1000 cubic centimetres:

(10 cm by 10 cm by 10 cm)

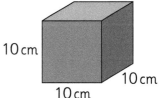
10 cm, 10 cm, 10 cm

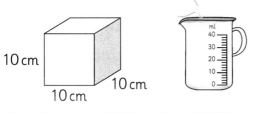
10 cm, 10 cm, 10 cm

One litre = 1000 ml = 1000 cm³

This container has a capacity of 1000 millilitres or one litre: (10 cm by 10 cm by 10 cm)

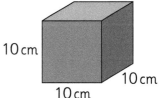

Let's practise

1) The following measurements have been given in millilitres. Convert each of the measurements into cubic centimetres:

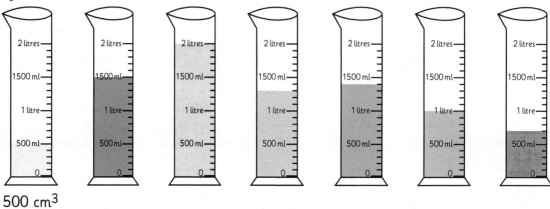

<u>500 cm³</u> _____ _____ _____ _____ _____ _____

2) The children have been asked to find the volume/capacity of each of the containers below in cubic centimetres and millilitres. They use cubic centimetres to fill them. They have found the measurements for the first container. Help them work out the measurements for the remaining containers:

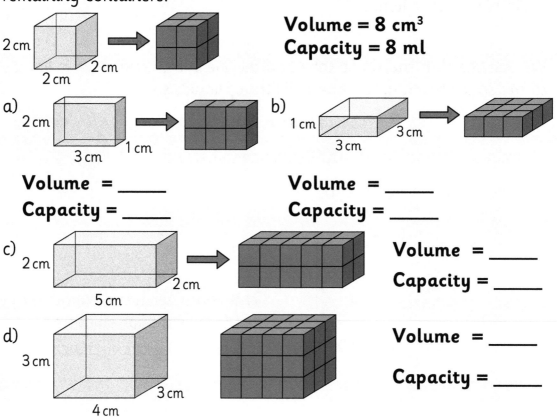

Volume = 8 cm³
Capacity = 8 ml

a) 2 cm, 3 cm, 1 cm

Volume = _____
Capacity = _____

b) 1 cm, 3 cm, 3 cm

Volume = _____
Capacity = _____

c) 2 cm, 5 cm, 2 cm

Volume = _____
Capacity = _____

d) 3 cm, 4 cm, 3 cm

Volume = _____
Capacity = _____

CHALLENGE!

Amman believes he has found a way to calculate the volume/capacity of each container without having to fill it with cubes:

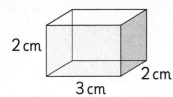

2 cm
3 cm
2 cm

I think it would take 12 cubes to fill this container so:

Volume = 12 cm³
Capacity = 12 ml

Can you explain how Amman has worked this out?

Use Amman's strategy to calculate the volume/capacity of the following containers:

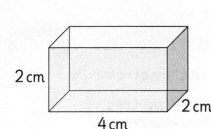

2 cm
4 cm
2 cm

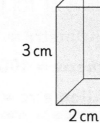

3 cm
2 cm
3 cm

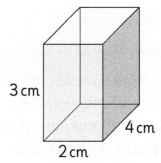

3 cm
2 cm
4 cm

How many different containers can you draw with a volume (capacity) of 24 cm³ (24 ml)?

9.5 Imperial measurement

Before we start

Convert the following:
- 2·4 metres to centimetres
- 3·5 litres to millilitres
- 4500 grams to kilograms

We are learning to use imperial measurements in everyday life.

We can use different measurement systems to measure length, mass and capacity.

Let's learn

We use the **metric system** for measuring:

- Length (mm, cm, m, km)
- Mass (g, kg)
- Capacity (ml, l)

Sometimes we use an older system of measurement called the imperial system. This uses different units of measurement:

The units for each type of measurement are linked by multiples of 10, for example:

1 centimetre = 10 millimetres

1 metre = 100 centimetres

1 kilometre = 1000 metres

Length	Mass	Capacity
1 foot is about 30 cm	1 pound is about 450 grams	1 pint is about half a litre
1 foot = 12 inches	1 pound = 16 ounces	1 pint = 20 fluid ounces
1 yard = 3 feet	1 stone = 14 pounds	1 gallon = 8 pints
1 mile = 1760 yards		

Let's practise

1 foot = 30 cm so I can calculate my height in metric units:

4 × 30 cm = 120 cm (or 1·2 m)

4 feet

1 foot
| 30 cm | 30 cm | 30 cm | 30 cm |
4 feet

1) Convert the following lengths into metric units:

    a) 3 feet   b) 10 feet   c) 7 feet   d) $\frac{1}{2}$ foot

I weigh exactly 5 stone. I wonder how many pounds that is?

1 stone equals 14 pounds. We can calculate your weight in pounds:

5 × 14 pounds = 70 pounds

1 stone
| 14 pounds | 14 pounds | 14 pounds | 14 pounds | 14 pounds |
5 stone

2) Convert the following weights into pounds:

    a) 6 stone   b) 9 stone   c) 20 stone

    d) $\frac{1}{2}$ stone   e) $7\frac{1}{2}$ stone

CHALLENGE!

Amman's bathtub holds 40 gallons of water.

Calculate the capacity of the bathtub in:

a) Pints   b) Fluid ounces

Investigate which is greater:

a) 1 metre or 1 yard   b) 1 kilometre or 1 mile

c) 4 gallons or 15 litres   d) 8 stone or 50 kilograms

9 Measurement

9.6 Calculating perimeter

We are learning to calculate the perimeter of regular shapes.

Before we start

What is the perimeter of both of these shapes?

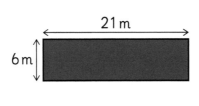

21 m

6 m

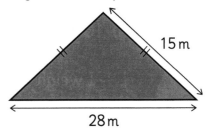

15 m

28 m

We can use decimal numbers to estimate and measure perimeter with greater accuracy.

Let's learn

We can estimate the perimeter of a shape by estimating the length of each side and adding them together:

I estimate the length of each side of this square to be 5 cm. I estimate the perimeter to be 20 cm:

4 × 5 cm = 20 cm

We can then measure accurately using a ruler:

The actual length of each side is 4·7 cm.
Therefore, the actual perimeter is 18·8 cm:

4 × 4·7 cm = 18·8 cm (or 188 mm)

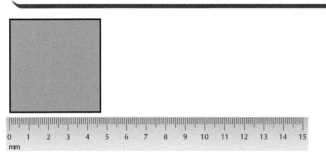

We can state the perimeter accurately using either centimetres or millimetres.

Let's practise

1) An architect is drawing plans for a new swimming complex. What is the perimeter of each section?

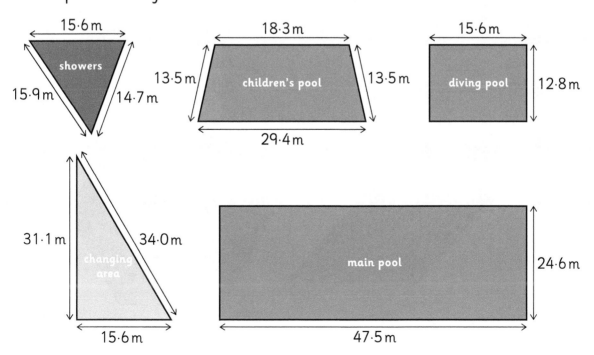

2) a) Estimate the perimeter of the following shapes.
   b) Use a ruler to measure them accurately.
   c) Calculate the difference.

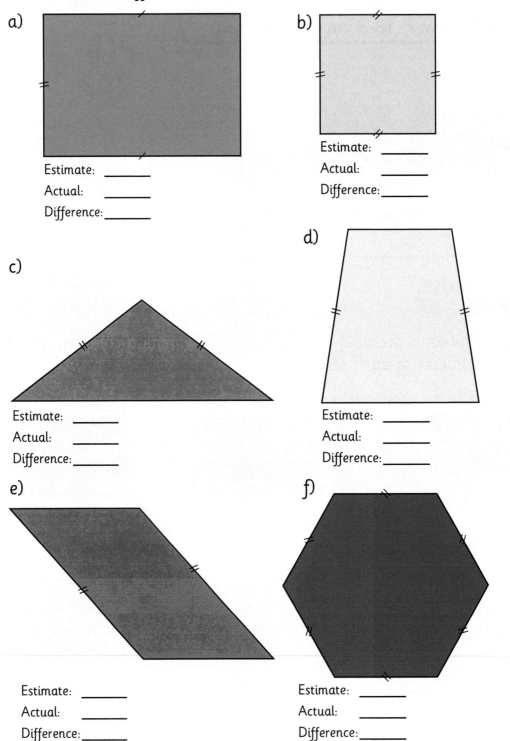

a)

Estimate: _____
Actual: _____
Difference: _____

b)

Estimate: _____
Actual: _____
Difference: _____

c)

Estimate: _____
Actual: _____
Difference: _____

d)

Estimate: _____
Actual: _____
Difference: _____

e)

Estimate: _____
Actual: _____
Difference: _____

f)

Estimate: _____
Actual: _____
Difference: _____

★ CHALLENGE! ....

Isla draws different shapes that all have a perimeter of 20 cm. She has written the lengths of some of the sides. Sketch each shape and fill in the missing measurements:

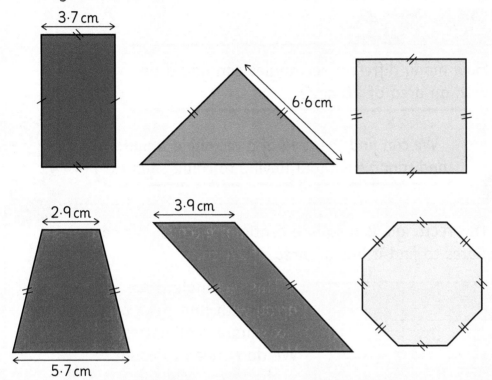

Draw three different shapes with a perimeter of 28·8 cm.

9.7 Calculating the area of regular shapes

We are learning to calculate the area of regular shapes.

Before we start

How many different rectangles can you draw with an area of 16 cm^2?

We can find the area of a rectangle, square and triangle without having to count squares.

Let's learn

Each box in this rectangle is a square centimetre (cm^2). We can count the squares to find it has an area of 24 cm^2:

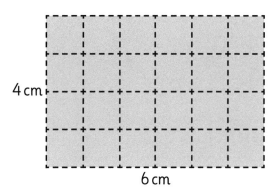

4 cm

6 cm

This rectangle is an array with four rows of six square centimetres. We don't need to count the boxes. We can work out the area as follows:

6 cm × 4 cm = 24 cm^2

We can use the formula, **Area = Length × Width** to calculate the area of a **rectangle** or **square**.

Isla cuts the rectangle in half to create two triangles that are exactly the same size:

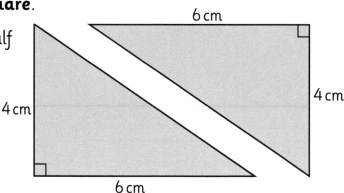

6 cm

4 cm

4 cm

6 cm

Each triangle is exactly half the size of the rectangle so their areas must be 12 cm². We can work out their area as follows:

$$\frac{1}{2} \times 4 \text{ cm} \times 6 \text{ cm} = 12 \text{ cm}^2$$

We can use the formula, **Area = $\frac{1}{2}$ × Base × Height** to calculate the area of a **triangle**.

Let's practise

1) Calculate the area of each of the following rectangles:

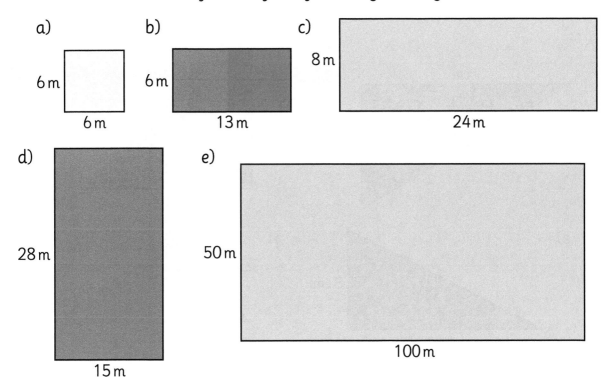

a) 6 m, 6 m

b) 6 m, 13 m

c) 8 m, 24 m

d) 28 m, 15 m

e) 50 m, 100 m

2) Calculate the area of each of the following (the first one has been done for you):

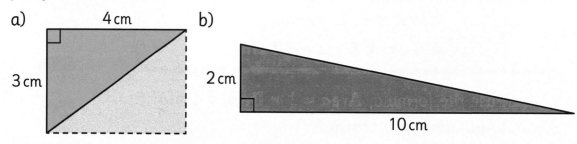

a) 4 cm

b)

3 cm

2 cm

10 cm

Area = $\frac{1}{2}$ × length × width

= $\frac{1}{2}$ × 4 cm × 3 cm

= 6 cm²

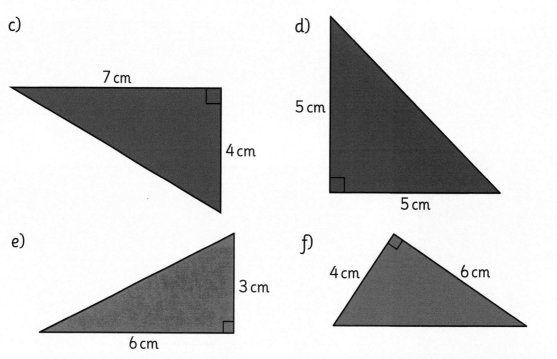

c)

7 cm

4 cm

d)

5 cm

5 cm

e)

3 cm

6 cm

f)

4 cm

6 cm

CHALLENGE!

Nuria has been challenged to find out how many rectangles and triangles she can draw with an area of 24 cm². She has attempted two of each so far:

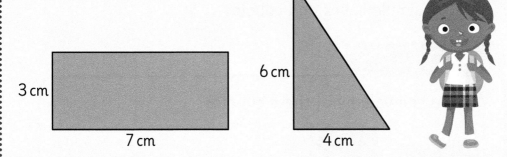

3 cm

7 cm

6 cm

4 cm

a) Calculate the area of each of Nuria's shapes. Has she managed to create a shape with an area of 24 cm²?

b) Help Nuria complete her challenge.

9.8 Calculating volume

We are learning to find the volume of simple cubes and cuboids.

Before we start

What is the volume of these cuboids?

Can you use blocks to make different shapes with the same volume as these?

We can work out the volume of a shape by counting layers.

Let's learn

The **volume** of this cube is one cubic centimetre (1 cm³):

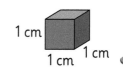

1 cm
1 cm   1 cm

This cuboid is made up of two cubic centimetres:

2 cm³

This cuboid is made up of three rows of 2 cm³:

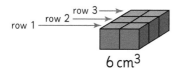

row 1   row 2   row 3

6 cm³

This cuboid is made up of three layers of 6 cm³:

If we want to work out the volume of any cuboid, we can identify how many cubic centimetres there are in a layer, then add up the total number of layers:

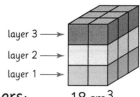

layer 3
layer 2
layer 1

18 cm³

6 cm³ + 6 cm³ + 6 cm³ = 18 cm³

1) The children are building cuboids using centimetre cubes. Each of them has drawn their plan for the first layer. How many cubic centimetres will they need altogether?

I want to make a cuboid with five layers

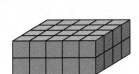

I want to make a cuboid with four layers

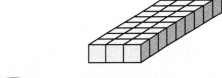

I want to make a cuboid with ten layers

I want to make a cuboid with six layers

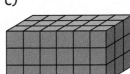

2) Calculate the volume of the following shapes:

a) 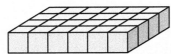   b)   c) 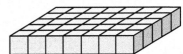   d)

CHALLENGE!

Nuria has been challenged to find as many cubes or cuboids as she can with a volume of 36 cm³. She thinks she has found one so far:

What is the volume of Nuria's attempt?

Help Nuria complete her challenge. How many different shapes can you find with a volume of 36 cm³?

Can you create a cube or cuboid with a volume of 45 cm³?

Mathematics, its impact on the world, past, present and future

10.1 Mathematical inventions and different number systems

> We are learning to discuss important mathematical discoveries and number systems.

Before we start

Discuss with a partner what you think a mathematician does?

Can you name any famous mathematical inventions that you use in school today?

Discuss with a partner the many ways numbers are used around us every day.

> Numbers and number systems are mathematical notations for representing numbers of a given set.

Let's learn

A number system is a way of recording and expressing numbers using digits, letters or symbols in a consistent way.

Around 3000 BC the ancient Egyptians used a system of numerical notation or 'hieroglyphs' based on a combination of the following shapes and marks:

I	1	I∩	11	∩∩∩	30
II	2	II∩	12	∩∩ ∩∩	40
III	3	III∩	13		
IIII	4	IIII∩	14	℗	100
IIIII	5	IIIII∩	15		
IIIII I	6	IIIII I∩	16	🔱	1000
IIIII II	7	IIIII II∩	17		
IIIII III	8	IIIII III∩	18		
IIIII IIII	9	IIIII IIII∩	19		
∩	10	∩∩	20		

1) Using the information from the table on the previous page, write down the following numbers using Egyptian numerals:

a) 19   b) 26   c) 31   d) 45
e) 331   f) 124   g) 55   h) 2018

2) Using the information from the tables on the previous page, write down the following facts about yourself in Egyptian numerals and ask a friend to check these:

a) Age   b) Date of birth
c) Shoe size   d) Number of people in your family
e) Pupils in your class

3) Copy and complete the following calculations using Egyptian numerals:

a) $\text{III}\cap + \cap\cap$ =

b) $\text{IIIII}\cap\cap - \text{III}$ =

c) $\text{IIIIII IIII}\cap\cap + \cap\cap\cap$ =

d) $\cap\cap - \text{IIIII}$ =
$\cap\cap$

⭐ **CHALLENGE!**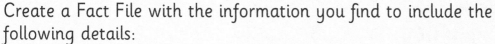

Let's Investigate...

James Gregory is a famous Scottish mathematician.

Work with a partner to research what he is famous for and what impact he has had on the world.

Create a Fact File with the information you find to include the following details:

- his name and date of birth (you could include his nickname!)
- where in Scotland he was from
- what he was best known for
- examples of his work (include pictures and diagrams)

11 Patterns and relationships

11.1 Exploring and extending number sequences

We are learning to identify and extend a pattern or sequence by explaining the rule.

Before we start

What is the next number in the following sequence?

1, 2, 4, 8, 16, 32,

Write the rule for the sequence in words.

A **number pattern** is a list of numbers that follow a certain sequence or pattern.

Let's learn

Using a table or a diagram can make it easier to identify a **number pattern**.

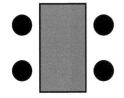

Each desk has 4 pupils

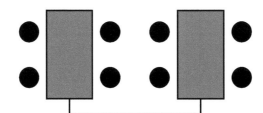

2 desks = 2 × 4 pupils

In the classroom, four pupils can sit at each desk.

Drawing a table can help you identify the pattern.

No. of desks (d)	1	2	3	4	5
No. of pupils (p)	4	8	12	?	?

For every new desk, the number of pupils increases by 4.

This can be written in words: **number of pupils = 4 × number of desks**

Or in symbol form: **p = 4 × d**

Therefore, for five desks there would be: **p = 4 × 5 = 20 pupils**

Let's practise

1) Each sponge cake has three candles.

   a) Copy and complete the table:

No. of sponges (s)	1	2	3	4	5
No. of candles (c)	3	6	9	?	?

   b) For each additional sponge cake, how many extra candles are required?

   c) Copy and complete the formula:

   Number of candles = _____ × number of sponge cakes

   d) Copy and complete the formula using symbols: **C = _____ × S**

   e) Use the formula to work out how many candles you would need for nine sponge cakes.

2) Each cardigan has five buttons.

   a) Copy and complete the table:

No. of cardigans (c)	1	2	3	4	5
No. of buttons (b)	5	10	15	?	?

   b) For each additional cardigan, how many extra buttons are required?

   c) Copy and complete the formula:

   Number of buttons = _____ × number of cardigans

   d) Copy and complete the formula using symbols: **B = _____ × C**

   e) Use the formula to work out how many buttons you would need for eight cardigans.

CHALLENGE!

Each day four school buses arrive at Abercraine Primary School with a total of 88 pupils.

   a) If the number of pupils on a school bus is always the same, copy and complete the following table by filling in the missing numbers:

No. of buses (b)	4	5	6	7	8
No. of pupils (p)	88	?	?	?	?

   b) Write a formula that connects the number of pupils (p) and the number of buses (b).

12 Expressions and equations

12.1 Solving equations using mathematical rules

We are learning to solve expressions and equations involving missing symbols, letters or numbers.

Before we start

For each number sentence below, find the unknown:

a) $166 = \boxed{} - 98$　　　　b) $133 - 47 = 193 - \boxed{}$

Using the symbols =, <, > and ≠, replace the ◆ to complete the following statements:

a) 10×5 ◆ $150 \div 3$　　　　b) 6×30 ◆ 18×10

c) $300 - 75$ ◆ 25×10　　　　d) 9×9 ◆ $200 - 50$

Equations are two number sentences that are connected by an equals sign.

Let's learn

An **equation** states that two things are equal.

It has an equals sign '=' like the following example:

$$12 \times 8 = 96$$

To solve any equation you need to ensure that the number statements on each side of the '=' sign balance and have the same total.

Sometimes we do not know one of the numbers. We can use a box or another shape to represent this unknown number.

Look at the following equation:

$$100 - \boxed{} = 50$$

Ask yourself '**what do I need to subtract from 100 to equal 50?**'

The answer is **50.**

We solve the equation by working out the unknown number.

1) For each sentence, find the unknown number.

    a) $3 \times \square = 15$ b) $16 \div \square = 4$ c) $\square \times 6 = 30$

    d) $\square \div 3 = 7$ e) $8 \times \square = 32$ f) $45 \div \square = 5$

2) For each sentence, find the unknown number.

    a) $3 \times \square = 18 - 6$ b) $20 \div \square = 1 + 3$

    c) $\square - 2 = 2 \times 9$ d) $2 + 2 = \square \div 4$

    e) $1 \times \square = 18 \div 6$ f) $16 \div \square = 2 \times 4$

3) Play the **Square Choice** game with a partner. Each player has five minutes to make as many number sentences as they can and write them in the columns provided.

- Pick two or three numbers from the grid and combine them with an unknown number box □ and any combination of × or ÷ to make the number sentence.

- After five minutes, each player tries to solve their opponent's number sentences. Score them:

    – For every number sentence solved with three numbers you score 3 points, for four numbers you score 4 points.

    – For number sentences that cannot be solved, you score 0 points.

64	18	6	36	64	4	70	**Player One**	**Player Two**
8	5	10	48	8	11	49		
81	72	7	30	54	21	90		
9	24	99	12	15	16	45		
14	6	35	3	63	9	7		
24	28	4	64	8	64	33		
5	10	20	60	27	11	2		

CHALLENGE!

Now work with larger numbers. For each number sentence, find the unknown.

a) $3 \times \boxed{} = 443 - 5$

b) $912 - 48 = \boxed{} \times 4$

c) $\boxed{} \times 3 = 97 \times 6$

d) $536 \div \boxed{} = 402 \div 6$

e) $2 \times \boxed{} = 352 \div 4$

f) $516 \div \boxed{} = 774 \div 3$

13 2D shapes and 3D objects

13.1 Naming and sorting polygons

We are learning to name and sort polygons by their properties.

Before we start

Finlay is looking at a shape with five sides, five vertices, one pair of parallel sides and two right angles. Name and draw Finlay's shape.

Polygons have straight sides. They are named for the number of sides that they have.

Let's learn

Triangles, squares, pentagons and hexagons are all polygons. Here are some other polygons:

A heptagon has seven straight sides

A nonagon has nine straight sides.

An octagon has eight straight sides.

A decagon has 10 straight sides.

A regular polygon has congruent sides (all the same length), and congruent angles (all the same size). An irregular polygon has sides that are of different lengths and/or has angles that are not all the same.

Let's practise

1) Name and draw the shape that matches the description.

a) This shape has:
   - Eight straight sides of equal length
   - Eight vertices

b) This shape has:
   - Six straight sides of different lengths
   - Six vertices

c) This shape has:
   - Four straight sides
   - Two pairs of parallel opposite sides

d) This shape has:
   - Three straight sides of the same length
   - Three vertices

2) Copy and complete the Venn diagram by writing the name of each shape in the correct place.

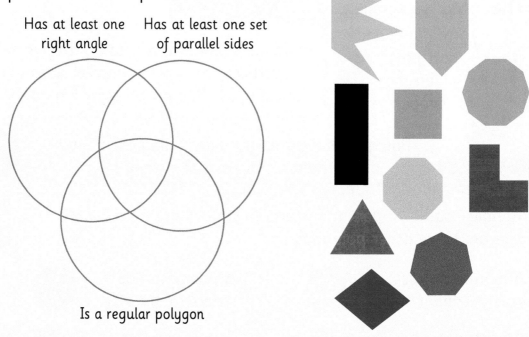

Has at least one right angle

Has at least one set of parallel sides

Is a regular polygon

3) Write two similarities and two differences for each pair of shapes. The first one has been done for you.

Shapes	Similarities	Differences
⬡ △	1 Have straight sides 2 No right angles	1 Decagon has 10 sides, triangle has only three 2 Triangle has three vertices, decagon has 10
⬢ ▭	1 2	1 2
⬠ ◼	1 2	1 2

CHALLENGE!

Draw at least two shapes in each set following the criteria. If shapes do not exist for the criteria, write 'shapes do not exist'.

	no pairs of parallel sides	exactly one pair of parallel sides	two pairs of parallel sides
no pairs of congruent sides			
more than one pair of congruent sides but not all sides are congruent			
all sides are congruent			

13.2 Describing and drawing circles

We are learning to describe and draw circles according to their properties.

Before we start

Which of these shapes are circles? Share the reasons for your choice with a partner.

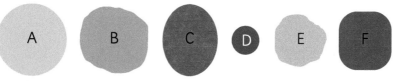

A B C D E F

Circles are a special kind of shape. They have properties that make them different from other shapes.

Let's learn

The distance around the circle is called the **circumference**.

Every point on the circumference is exactly the same distance from the **centre**. This distance is called the **radius**.

The distance across the circle through the centre is called the **diameter**.

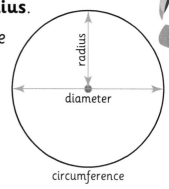

radius

diameter

circumference

Did you know that thousands of years ago, people worked out that the circumference of every circle is *approximately* **three times** the length of the diameter?

To draw a circle, use a pair of compasses. Separate the ends so that the distance between them is exactly the same as the radius of the circle you want to draw. Press the pointed end onto the paper and spin the drawing end round, holding the compasses at the tip.

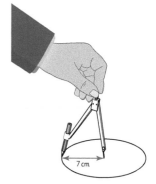

7 cm

Let's practise

1) a) Copy and complete the table by carefully measuring

   i) the radius and

   ii) the diameter of these circles to the nearest millimetre.

   b) Write down a rule to find the diameter when you know the radius.

Circle	Radius (mm)	Diameter (mm)
A	15	30
B		
C		
D		
E		

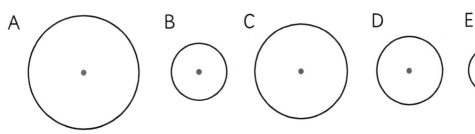

A   B   C   D   E

   c) Use your measurement of the diameter to estimate the circumference.

2) a) Calculate the missing measurements in the following table:

Circle	Radius	Diameter	Circumference (approximately)
F	14 cm		
G		50 mm	
H			36 cm
I		9 cm	
J			126 mm

   b) Using a pair of compasses, draw circles F to J.

3) This design is made of circles, so the circumference of one circle passes through the centre of the next. Copy and continue the pattern, using a radius of 5 cm for all your circles.

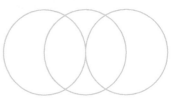

⭐ **CHALLENGE!** ....

Create your own design using

a) circles of diameter 5 cm.   b) circles with different diameters.

13 2D shapes and 3D objects

13.3 Constructing 3D objects

We are learning to make skeletons of 3D objects.

Before we start

This is one face of a 3D object.
Which of the following can the 3D shape NOT be?
a) A cube
b) A cuboid
c) A cylinder
d) A pyramid
e) A triangular prism

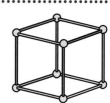

Constructing skeletons of 3D objects helps us to understand their properties.

Let's learn

You can make a skeleton of a cube using 12 straws of the same length and eight blobs of modelling clay. Work with a partner for the activities on these pages, and to answer the questions.

Let's practise

1) You will need solid versions of each of these 3D objects.

a)

b)

c)

Use straws and modelling clay to show the edges and vertices of these shapes.
a) Lay the straws along the edges. Trim them to the correct length with scissors.
b) Place a blob of modelling clay at the vertices.
c) Press the straws into the clay.

2) a) Follow the instructions in the table to make skeletons of these 3D objects, using two different lengths of straw.

Long straws	Short straws	Blobs of modelling clay	Name of 3D object
0	12	8	Cube
4	8	8	Cuboid
3	3	4	Triangle-based pyramid
4	4	5	Tall square-based pyramid
0	8	5	Short square-based pyramid
3	6	6	Triangular prism

b) Look at the objects you have made and answer the questions.
   i) Which have more than three edges meeting at a vertex?
   ii) Which have more than two triangular faces?
   iii) Which have more than three rectangular faces?

3) Explore the strength and stability of the 3D objects you constructed in question 2.
   a) Which objects can you stack a cube on top of?
   b) What happens if you press down on the top?
   c) What conclusions can you draw about the 2D shapes within the 3D objects?

CHALLENGE!

Work with a partner. Build the three skeletons of cubes using interlocking cubes.

Count the number of cubes in each skeleton and complete the table:

Skeleton of cube	Total number of cubes
3 by 3 by 3	
4 by 4 by 4	
5 by 5 by 5	

13 2D shapes and 3D objects

13.4 Constructing nets

We are learning to construct nets of cubes, cuboids and pyramids.

Before we start

Copy and complete the table:

3D object	Faces	Vertices	Edges
Cube	6		
Cuboid		8	
Triangle-based pyramid			
Square-based pyramid			

A net is a flattened 3D object. It can be made into the object by folding.

Let's learn

This is the net of a square-based pyramid.

The triangles can be folded upwards along the edges of the square, making a three-dimensional square-based pyramid.

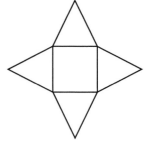

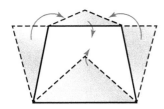 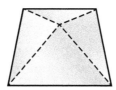

Let's practise

1) This is the net of a cube.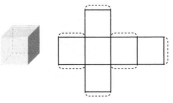

   Copy the net carefully, using a ruler to make
   the edges 4 cm (squared paper will help).
   Cut out your net, making sure to include the tabs.
   Fold each solid line first, then fold the net up to
   make a cube. Apply glue to the tabs to stick the cube together.

2) Here are the nets of some 3D objects.

   a)    b) 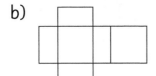   c)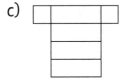

   a) Predict what object each net will make when it is folded.
   b) Copy the nets carefully onto squared paper, using a ruler. Cut them
      out along the outside edges. Fold them to check your predictions.

3) Which of these shapes are actual nets of cubes? Copy and fold them
   to check.

   a) 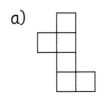   b) 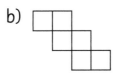   c)   d)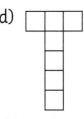

   e)   f) 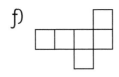   g)   h)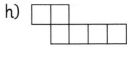

⭐ **CHALLENGE!** ....

This is just one possible net of a square-based pyramid.
There are five other possible nets: can you find them?
Draw them, then cut them out and fold them to check.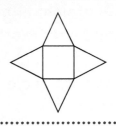

14 Angles, symmetry and transformation

14.1 Identifying and sorting angles

We are learning to identify and sort right, acute, obtuse, straight and reflex angles.

Before we start

Spot the errors.

Acute angles 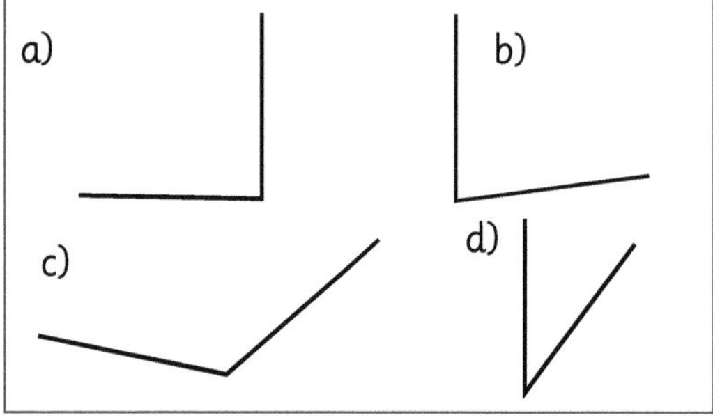	Obtuse angles 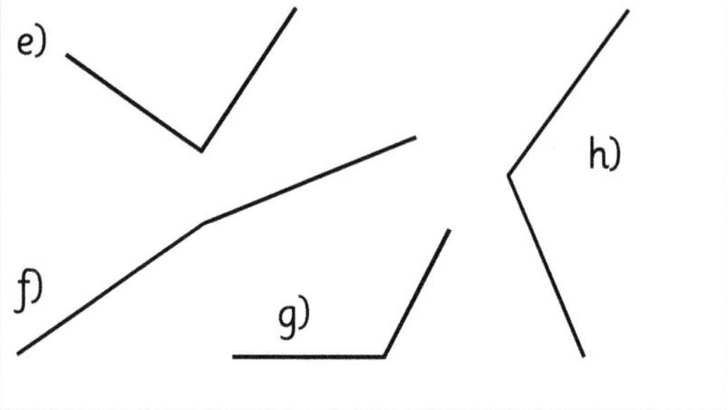
a) b)	e) h)
c) d)	f) g)

Let's learn

Angles can be categorised by their size, up to 360° (a full turn).

A right angle is a quarter turn. It measures exactly 90°.	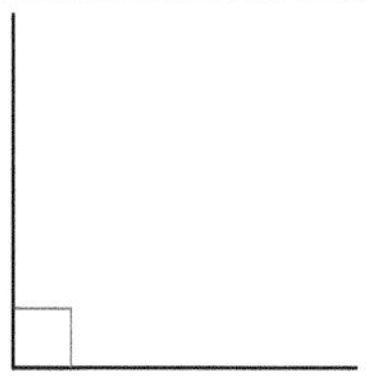
An acute angle is smaller than a right angle. It measures less than 90°.	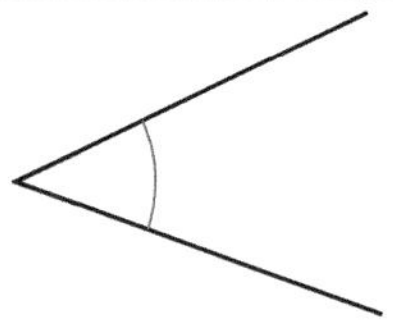
An obtuse angle is bigger than a right angle. It measures more than 90° but less than 180°.	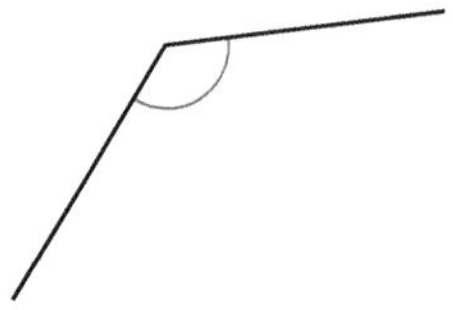
A straight angle is a half turn. It measures exactly 180°.	
A reflex angle is bigger than a straight angle. It measures more than 180°.	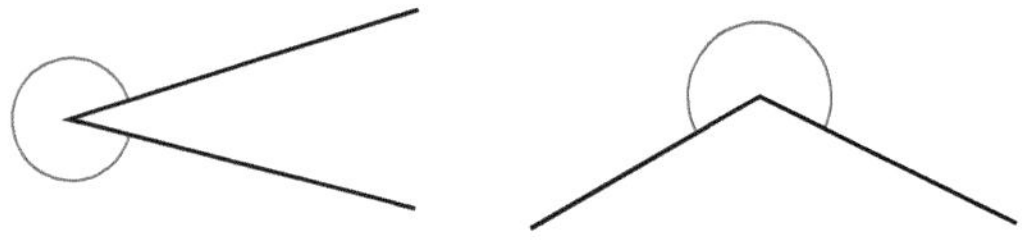

Let's practise

1) Copy and complete the table to sort the angles. One has been done for you.

Acute	Obtuse	Reflex
a		

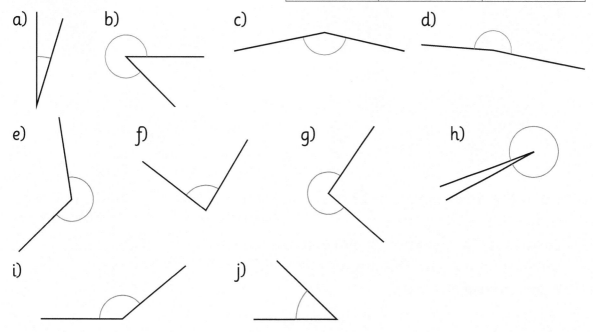

a) b) c) d)

e) f) g) h)

i) j)

2)

Number of reflex angles	Number of acute angles	Number of obtuse angles

a) Look at the diagram, then copy and complete the table.

b) Use a ruler to create your own diagram. Challenge a friend to find all the different types of angles.

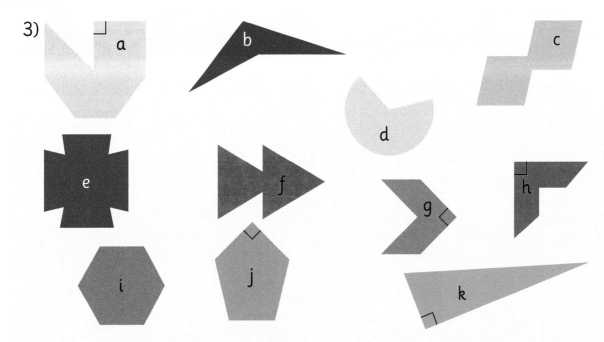

Copy the Venn diagram. Sort the shapes by writing the letter of each shape in the correct place in the Venn diagram. The angles are those **inside** the shape.

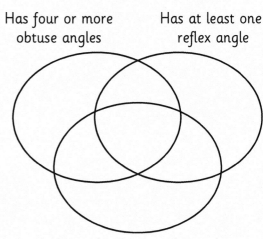

Has four or more obtuse angles

Has at least one reflex angle

Has at least one right angle

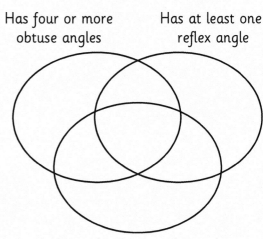

⭐ **CHALLENGE!**

Amman claims that it is impossible to have an obtuse angle, a reflex angle and an acute angle greater than 45°, all within one full turn. Is he right? Justify your answer.

14.2 Measuring and drawing angles

We are learning to accurately measure angles up to 360° and to draw angles up to 180°.

Before we start

Finlay says this is an acute angle. Isla says it is a reflex angle. Who is right? Explain your thinking.

We can use a protractor to measure exactly how many degrees there are in an angle, and also to draw angles accurately.

Let's learn

To measure reflex angles, we use the knowledge that a full turn is 360°.

First, measure the part that is opposite the reflex angle. Then subtract it from 360° to give the reflex angle.

This angle measures 40°

For example, 360° – 40° = 320° so the reflex angle marked here in red measures 320°.

To draw angles, we use a protractor. Follow these steps:

Draw a horizontal line (AB).

Draw a dot at point B. This is the vertex.

Line up the 'zero line' of the protractor with line AB.

Make sure the vertex of the angle (B) is on the centre point of the protractor

Make a mark on the scale the size of your angle (C).

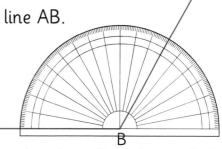

Remove the protractor and use a ruler to draw a line from B to the mark C.

14

1) Measure these angles accurately:

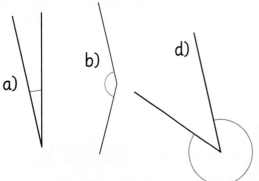
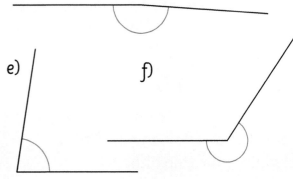

a) b) c) d) e) f)

2) Nuria has got lost in a maze. Measure the angles she needs to turn to get home.

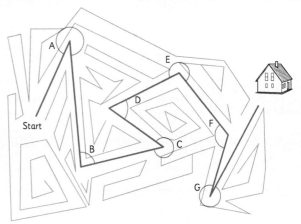

3) Accurately draw angles measuring:

a) 69°   b) 128°   c) 31°
d) 47°   e) 104°   f) 162°

CHALLENGE!

Draw a horizontal line in your jotter.

a) Estimate, then draw a line where you think the angle 30° would be. Check how close you are by measuring.

Repeat for the folowing angles:

b) 45°   c) 110°   d) 240°

Now draw some angles for a partner to estimate, then measure.

14 Angles, symmetry and transformation

14.3 Finding missing angles

We are learning to calculate missing angles.

Before we start

Estimate, then measure this angle.

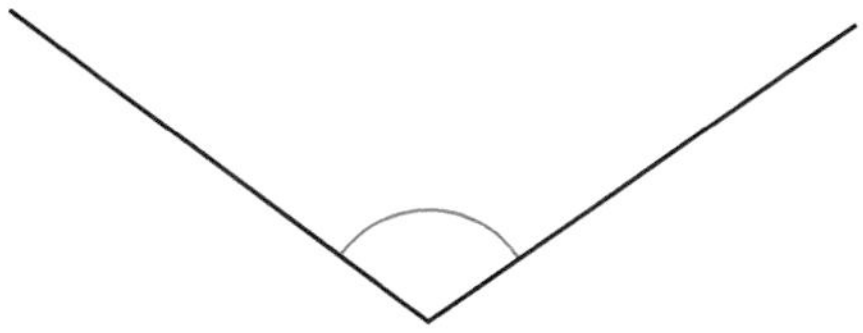

We don't always need to measure an angle. Sometimes we can calculate how many degrees it is instead.

Let's learn

When two angles add up to a right angle (90°) we call them **complementary**.

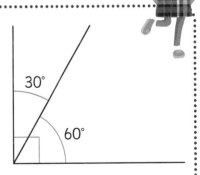

30°

60°

When two angles add up to a straight angle (180°) we call them **supplementary**.

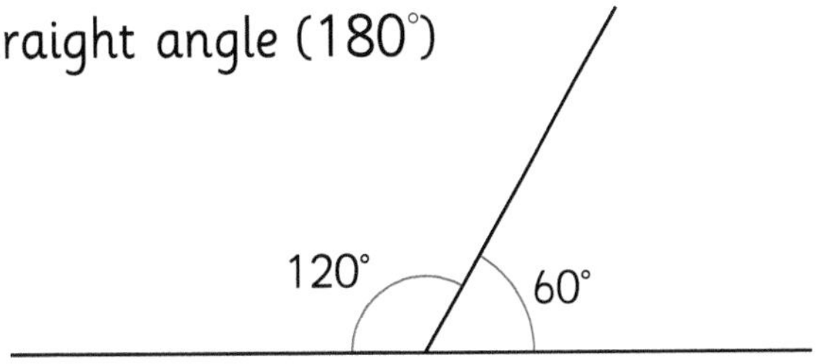

120° 60°

14

Let's practise

1) Calculate the missing complementary angles.

a)

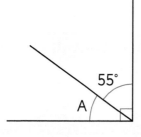

55° A

b)

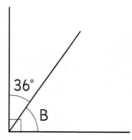

36° B

c)

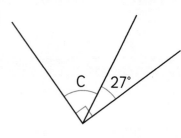

C 27°

d)

D 81°

e)

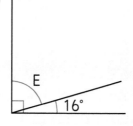

E 16°

2) Calculate the missing supplementary angles.

a)

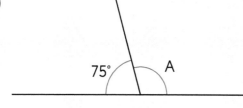

75° A

b)

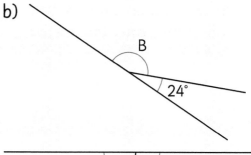

B 24°

c)

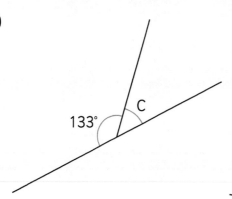

133° C

d)

86° D

e)

E 33°

3) Measure angle A with a protractor, then calculate to find angle B.

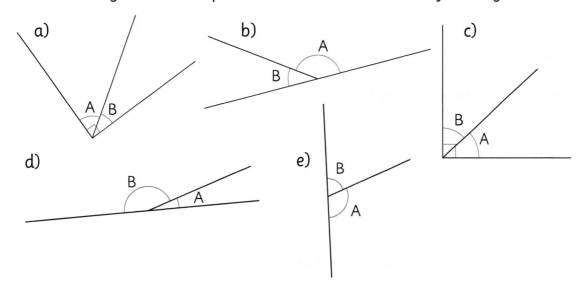

a) b) c)

d) e)

CHALLENGE!

a) A pair of complementary angles, A and B have a difference of 12°. Angle A is smaller than angle B. Calculate A and B.

b) In a pair of supplementary angles, C and D, angle C measures five times angle D. Calculate C and D.

c) Can you think of a story that will help you remember the size of complementary and supplementary angles?

14 Angles, symmetry and transformation

14.4 Locating objects using bearings

We are learning to locate objects on a map or plan using bearings.

Before we start

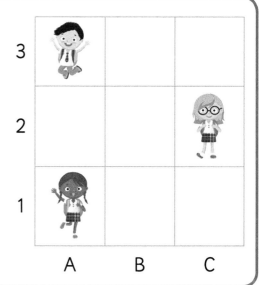

a) Amman is hiding. Finlay is one square north-west of Amman. Which square is Amman hiding in?

b) Finlay wants to join Isla. He has two moves. He travels one square east. What is his next move?

Sailors and airline pilots use three-figure bearings to make sure that their ship or plane is travelling in precisely the right direction. Each three-figure bearing describes a unique direction.

Let's learn

An eight-point compass can locate objects or places in eight directions. Bearings locate objects or places in any direction. Bearings are measured in degrees, like angles.

Bearings are always measured in the same way:

1) Start at north.

2) Measure clockwise.

3) If the bearing measures more than 180°, measure from south and add 180°.

4) Use three figures. If a bearing measures less than 100°, include a zero at the beginning.

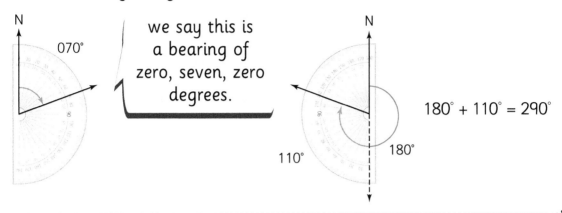

we say this is a bearing of zero, seven, zero degrees.

180° + 110° = 290°

Let's practise

1) Here are the three-figure bearings for north, south, east and west.

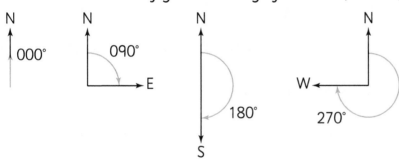

Calculate the three-figure bearings for:
a) north-east   b) south-west
c) south-east   d) north-west

2) Look at this plan. Copy and complete the table.

From	Compass direction	Bearing	To
Circle			Rhombus
Kite	North-east		
Rhombus		000°	
Trapezium			Triangle
Square		315°	
Star			Pentagon

3) Amman and his dad are standing on the peak of Cairn Gorm Mountain in Cairngorm National Park.

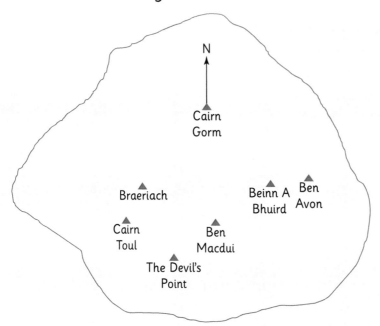

Amman uses a protractor to measure the bearings of the different mountains from the top of Cairn Gorm. Write down the name of the mountain that is on a bearing of:

a) 215°   b) 140°   c) 175°   d) 220°

CHALLENGE!

A ship is sailing on a bearing of 040°.

It sees a flare and decides to change course to head towards the flare and investigate.

What bearing does the ship need to take now?

What bearing would the ship need to take to return from the flare to the start point?

What bearing would it then need to take to face the same direction as it is in the diagram?

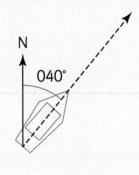

14.5 Reading coordinates

We are learning to read the coordinates on a coordinate grid.

Before we start

Copy the grid. Plot the points (4,3) (6,5) (2,5) (0,3) and (4,3). Join them up as you go. What shape have you made?

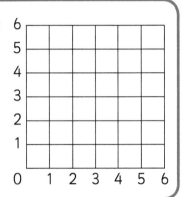

Coordinates describe exactly where a point is on a graph or grid. We can read the coordinates of a point by counting horizontally, then vertically.

Let's learn

To read the coordinates, first count horizontally from the *origin*, the point (0, 0). We count along two so we write the coordinate (2,)

Then count vertically. This gives the second coordinate (2, 4).

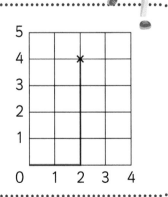

14

1) Write down the coordinates of the vertices of each shape.

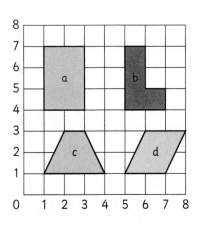

2) a) Write down in the correct order, starting at the origin, the coordinates of the points that make the letter N on the grid.

   b) Copy the grid and plot points to make these letters. Write down the coordinates in the correct order.
   i) M   ii) Z   iii) W

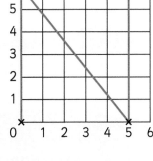

3) a) Write down the coordinates of the vertices of this shape, clockwise starting at (1,6).

   b) The shape is reflected in the dotted line. Starting at (2,3), write down, clockwise, the coordinates of the vertices of its reflection. (You may wish to copy the grid and plot the points first.)

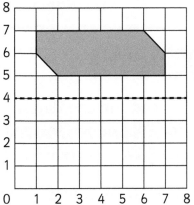

⭐ **CHALLENGE!** ....

The coordinates of point A are (3, 7). Write down the coordinates of all the other vertices.

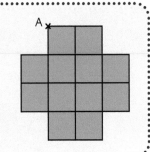

14 Angles, symmetry and transformation

14.6 Line symmetry

We are learning to identify up to two lines of symmetry in shapes.

Before we start

Isla says this shape has one horizontal line of symmetry.

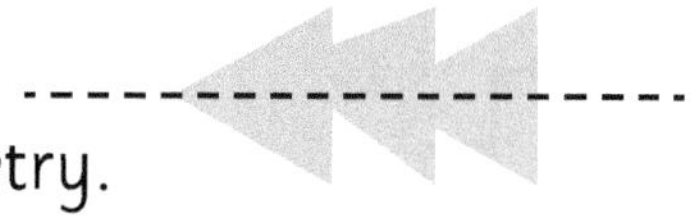

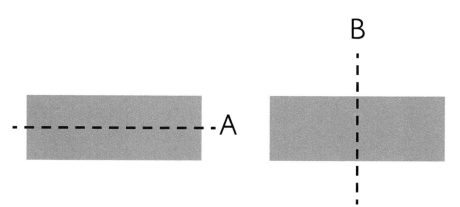

Finlay says the shape is not symmetrical. Who is right? Explain your thinking.

Some shapes have more than one line of symmetry.

Let's learn

A rectangle has two lines of symmetry. If you place a mirror on line A, the reflection is exactly the same as without the mirror. If you place a mirror on line B, the reflection shows the same shape as without the mirror.

This can also be shown by folding. Take a rectangular piece of paper and fold it exactly in half, along each line of symmetry shown above. The corners and edges will meet exactly.

1) These shapes all have two lines of symmetry.

Copy them into your jotter. Draw in both lines of symmetry for each shape.

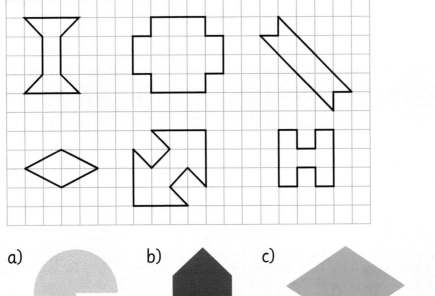

2) a)   b)   c)   d)

e)   f)   g)   h)

i)   j)   k)

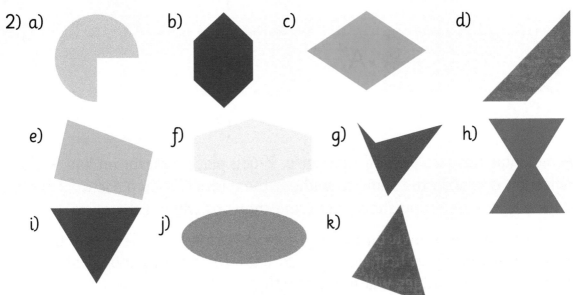

Use a mirror to find the lines of symmetry in these shapes.

Copy the table, then sort the shapes by writing the letters in the correct column.

No lines of symmetry	One line of symmetry	Two lines of symmetry

3) Copy these designs into your jotter.

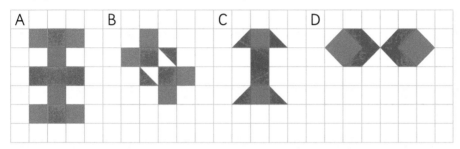

a) Draw the lines of symmetry for each design.

b) Which is the odd one out?

CHALLENGE!

Which of these designs is correctly reflected in both lines of symmetry?

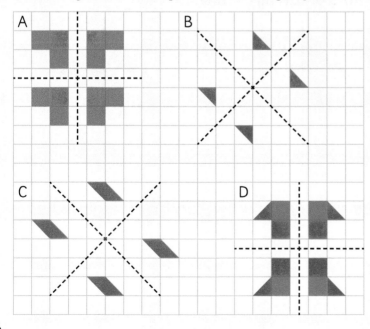

14.7 Symmetrical pictures and diagrams

We are learning to complete symmetrical pictures by reflecting in two lines of symmetry.

Before we start

Isla says she has completed this shape by reflecting it in the mirror line. Do you agree? Explain your thinking.

When we complete a shape so it has two lines of symmetry, we reflect it first in one mirror line, then the second.

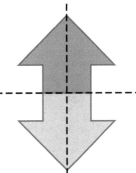

Let's learn

To complete this shape, first it was reflected in the vertical mirror line. Then it was reflected in the horizontal mirror line.

Let's practise

1) Copy these shapes into your jotter, then complete them by reflecting them first in one mirror line, then the second.

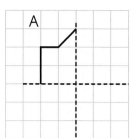 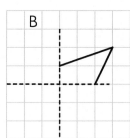 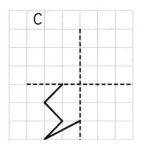 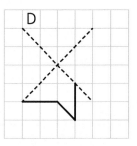

2) This quilt design is only one quarter made. Reflect the pattern in the mirror lines to complete the symmetrical design.

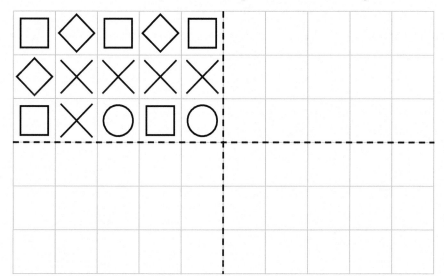

3) Complete the design for each square tile by reflecting the square pattern in both lines of symmetry.

a) 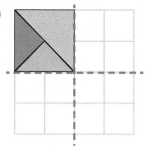   b) 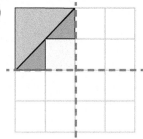   c)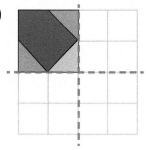

4) a) Create and complete your own design for a square tile. Make sure you include two lines of symmetry.

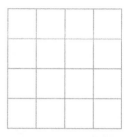

   b) Describe the steps of reflection that you took to make the design.

CHALLENGE!

Reflect the pattern in both lines of symmetry.

Add colour to your design to highlight its symmetry.

Reflect the complete design vertically onto the other half of the square dot grid.

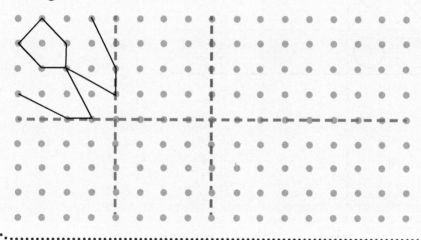

14.8 Reading scale maps

> We are learning to interpret simple maps or plans with a given scale.

Before we start

A map shows two towns. They are 7 cm apart on the map. If the scale of the map is 1 cm for every 5 km, what is the real-life distance:

a) 350 km

b) 7 km

c) 35 km

d) 5 km

> The scale tells you how much bigger everything on a map or plan is in real life.

Let's learn

Maps and plans represent real-life distances or measurements, scaled down to fit on a page. If we know the scale, we can calculate the real-life distance accurately.

On this map of Sunnytown, the distance between A and B is 5 cm.

The actual distance between A and B is 5 km.

Every 1 cm on the map represents 1 km in real life. We write the scale as **1 cm : 1 km**

This scale looks like this:

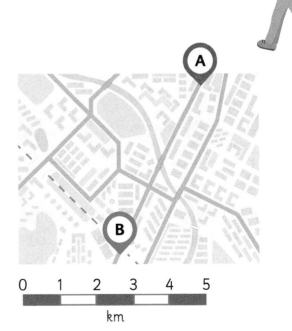

14

Let's practise

1) Here are some scales taken from maps.

   a) For each one, write down the scale in the form **1 cm :____**
   b) Calculate how far 8 cm on each map would be in real life.

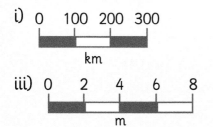

   i) 0 100 200 300 — km

   ii) 0 5 10 15 20 — km

   iii) 0 2 4 6 8 — m

   iv) 0 150 300 450 — km

2) Here is a city plan of Happyville.

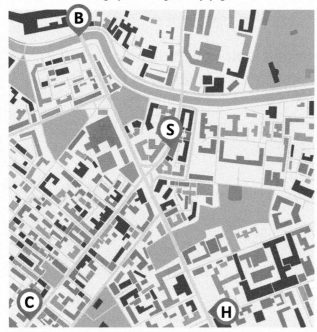

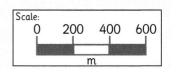

Scale: 0 200 400 600 m

   a) What is the scale of the plan? Write it in the form **1 cm :____**
   b) Amman is on the bridge (B). How far is it to the hospital (H)?
      Measure the distance on the map to the nearest centimetre and
      write it down, then calculate the distance in real life.
   c) Nuria's class is going from school (S) to the church (C). Measure
      and write down the distance on the map to the nearest centimetre
      then calculate the distance in real life, **in kilometres.**
   d) Finlay goes for a walk. He walks for three kilometres. How far is
      three kilometres on the map?

3)

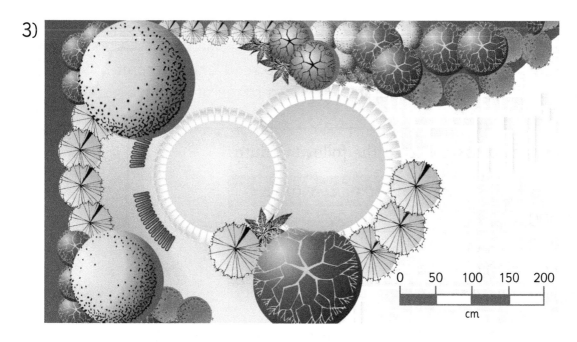

0   50   100   150   200

cm

A garden designer has made a plan of her design for a client.

a) What scale has she used?

b) What are the dimensions (length and breadth) of the client's garden to the nearest metre?

c) What will be the size in real life of the diameter of
   i) the smaller lawn?   ii) the larger lawn?
   Give your answers in metres.

d) The designer has bought a bench which is 2 m long and 0·5 m wide. What size does she need to draw it on the plan?

⭐ **CHALLENGE!**

Draw a plan of your classroom, or another room in your school. You will need to measure the length and breadth of the room, and the dimensions of the furniture. Decide on a suitable scale so your plan can fit on one page.

Write your scale in the corner of your plan.

15 Data handling and analysis

15.1 Working with a range of graphs

We are learning to read and interpret information in a range of graphs.

Before we start

A leisure club has collected the following data:

Time	Number of customers
17:00 – 18:00	27
18:00 – 19:00	43
19:00 – 20:00	48
20:00 – 21:00	15

What type of graph should they use to display this data? Explain your thinking.

Bar graphs, pie charts and line graphs are just some of the ways in which data can be represented.

Let's learn

Newspapers, science reports and textbooks often use other types of graph to represent information.

Double bar and double line graphs show two sets of related data on the same graph. They use different colours to show each set of data.

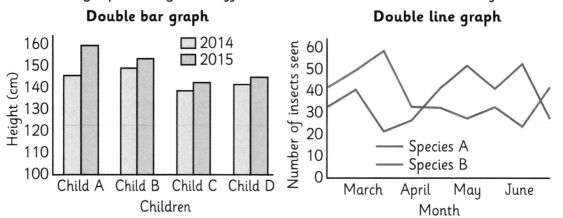

Let's practise

1) A school holds a quiz each year involving two teams. The results are shown in the double bar graph. Answer the questions that follow.

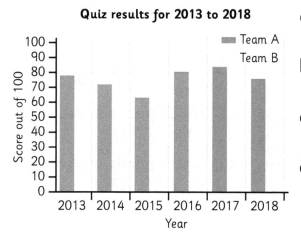

Quiz results for 2013 to 2018

a) What score did Team B achieve in 2018?

b) In which years did Team A score 75 or more?

c) In which years did Team B score over 80?

d) Which year produced the biggest difference in the scores of the two teams?

2) The double line graph shows the number of insects of two species, counted for the period January to July. Answer the questions that follow.

Number of insects observed from January to July

a) In February, how many more of Species A than Species B were counted?

b) In April, how many more of Species B than Species A were counted?

c) Which month showed the greatest difference in species numbers?

d) How many of species A were observed altogether from January to April?

e) How many of species B were observed altogether from April to July?

3) Study the 3D double bar graph and answer the questions that follow.

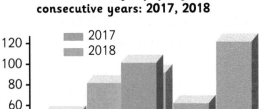

Maths test scores for pupils for two consecutive years: 2017, 2018

2017
2018

Total scores / Pupils

a) Which pupil shows the most progress from 2017 to 2018?

b) Which pupil performed worse in 2018 than in 2017?

c) Which pupils scored over 20 in 2017?

d) Which pupils scored over 60 in 2018?

CHALLENGE!

Two lorries set off from the depot to deliver furniture to the same store, which is 180 km from the depot.

The line graph shows their journeys.

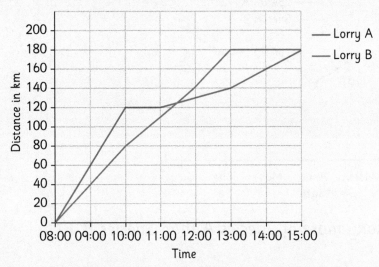

Lorry A
Lorry B

a) Which lorry arrived at the store first?

b) Which driver stopped during the journey?

c) At approximately what time were both lorries the same distance from the depot?

d) How far did Lorry B drive between 11:00 and 13:00?

15.2 Using pie charts

We are learning to read and interpret data from pie charts and to draw pie charts.

Before we start

24 children are asked their favourite type of holiday. 12 of them say they like activity holidays best. The data is displayed in a pie chart.
Which of these pie charts is not correct?

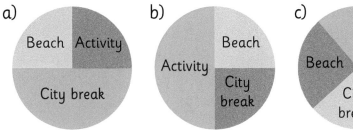

a)

b)

c)

We use the fact that there are 360° in a full turn to interpret pie charts.

Let's learn

36 pupils were asked which was their favourite fruit from the tuck shop.

There are 36 pupils in the whole, so 36 pupils measure 360° on the pie chart.

360 ÷ 36 = 10, so one pupil is represented on the pie chart by 10°.

10° = one pupil

Here is a pie chart displaying all the data.

The yellow sector measures 120°. 120 ÷ 10 = 12, so 12 pupils chose banana as their favourite fruit.

The green sector measures 150° so 150 ÷ 10 = 15 pupils chose apple as their favourite fruit.

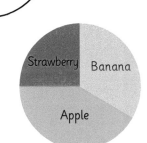

The red sector measures 90°, so nine pupils liked strawberries best.

Remember to check that we have found the correct number of pupils by adding: 12 + 15 + 9 = 36.

Let's practise

1) 36 people were asked where they went on their summer holiday.

   Here is a pie chart of their responses. Measure each section carefully using a protractor to answer the questions.

   a) Which was the most popular holiday destination?
   b) How many people went to Scotland on holiday?
   c) How many people did not go away?
   d) How many more people went to Spain than France?

Holiday destinations

2) A pet shop owner conducted a survey of 72 dog owners to find out which dog food they feed their dog. The results are shown in the pie chart.

   Hint: 360 ÷ 72 = 5°

   a) Which is the most popular dog food?
   b) How many dog owners feed their dog Bouncer's Original?
   c) How many more dog owners prefer Woofit Mix to Sir Scoffalot?
   d) How many dog owners feed their dogs either Doggo Yummies or Meaty Marvel?

Favourite dog food

3) Here is a frequency table, showing how many hours of TV a sample of 10-year-olds watched on school days.

Hours spent watching TV on school days	Number of 10-year-olds	Angle in pie chart
0	7	
Less than an hour	11	
1–2 hours	13	
2–3 hours	4	
More than 3 hours	1	
Total	36	360°

a) Copy and complete the table.
b) Draw a pie chart displaying the data. You will need to use a protractor to measure the angles.

CHALLENGE!

A chef has created five new recipes. She wants to know which is the most popular. In a blind taste test, she labels the recipes A, B, C, D and E and tries them out on some customers.

The results are shown in the pie chart.

60 people in the survey liked recipe D.

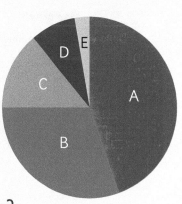

a) How many people were surveyed altogether?
b) Which was the most popular recipe?
c) How many people liked the most popular recipe?

15 Data handling and analysis

15.3 Creating and interpreting graphs

We are learning to create graphs from data gathered as part of an inquiry and use them to interpret the data.

Before we start

Amman says 'I wonder how tall my bean plant will grow each day for the next week.' Create a table for Amman to record the results.

When we conduct an inquiry, graphs can help us to communicate what we have found out.

Let's learn

Graphs help us to communicate information. This is why we see graphs in television news reports, newspapers and magazines.

We need to choose a type of graph to match the data that we have collected.

It is best to show category data in a bar graph or a picture graph.

It is best to show time series data in a line graph.

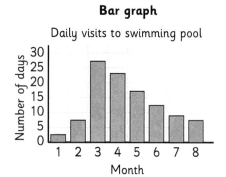

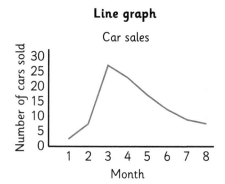

Let's practise

1) Which type of graph would you choose to show each set of data?

a) Big question: 'I wonder how many other countries students have visited this year.'

Number of countries	Number of students
0	6
1	8
2	14
3	7
4 or more	3

b) Big question: 'I wonder how the height of a ramp affects the distance a toy car travels.'

Height of ramp (cm)	Distance travelled (cm)
10	35
20	43
30	50
40	56
50	61

c) Big question: 'I wonder what our favourite colours are.'

Colour	Number of students (tally marks)
Red |卌 |||
Blue	卌
Green	|||
Yellow	卌
Purple		

2) Are the statements about the graph on the right true or false?

a) 'I notice that chess is more popular than painting.'

b) 'I notice that three of the bars are close to the same height, and that one bar is a lot smaller.'

c) 'I notice that the least popular hobby is reading.'

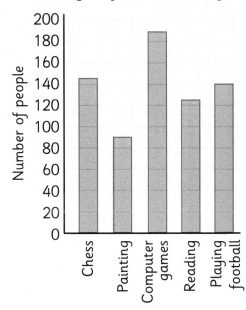

What is your favourite hobby?

d) 'I notice that computer games are over three times more popular than painting.'

e) 'I notice that the most popular hobby is computer games.'

3) a) Write two 'I notice ...' statements about this line graph.

b) Look at the bean plant graph again. Use information from the graph to complete the statements.

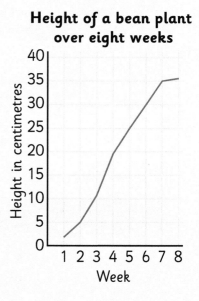

Height of a bean plant over eight weeks

i) I notice that the line goes up for most of the graph. This means ...

ii) I notice that the line starts to level out after week 7. This means ...

iii) I notice that between weeks 4 and 7 the line goes up by a similar amount each week. This means ...

iv) Write three 'I notice statements ...' for the data you are working with in your statistical inquiry.

CHALLENGE!

Plan your own inquiry.

a) Think of a big question. Start with 'I wonder ...'.

b) Make a table to collect your data. Give it suitable headings, including units of measurement if needed.

c) Choose an appropriate graph to display your data.

d) Write as many 'I notice' statements as you can about your graph.

15.4 Drawing conclusions from graphs

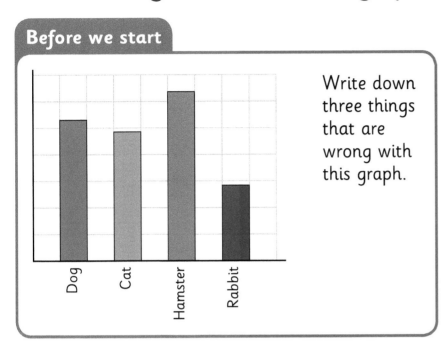

Write down three things that are wrong with this graph.

We are learning to draw conclusions from the data displayed in a graph.

When we conduct an inquiry, we are looking for answers to our 'big question'. Interpreting graphs helps us to find the answers to our big question.

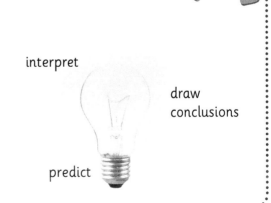

Let's learn

At the end of an inquiry, we need to make clear conclusions that answer our original 'big question'.

We can then extend an inquiry by writing 'I wonder' questions. We can make predictions about our new inquiry by writing 'I think' statements.

interpret

draw conclusions

predict

1) The bar graph shows the data from an inquiry. Look at the statements in the table. Are they appropriate for this inquiry? Write yes or no.

   a) I found that people had lots of hobbies but the most popular was reading.

   b) I wonder why so many people prefer computer games.

   c) I wonder why so many people don't like football.

   d) I think most people have lots of hobbies.

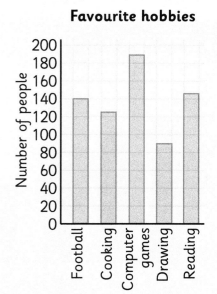

Favourite hobbies

2) The line graph shows the data from an inquiry. Look at the statements in the table. Are they appropriate for this inquiry? Copy and complete the table, placing a tick in the correct column.

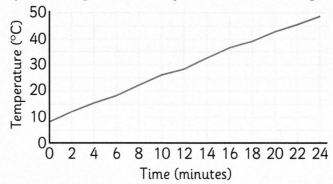

Temperature of water in a pan that is heated gradually

Statement	Yes	No
'I found that the water steadily increased in temperature.'		
'I wonder if we would get similar results if the experiment began with warmer water.'		
'I wonder why water cools so quickly.'		
'I think someone spilled some of the water.'		

3) This graph shows the average temperatures over a year in Glasgow and London.

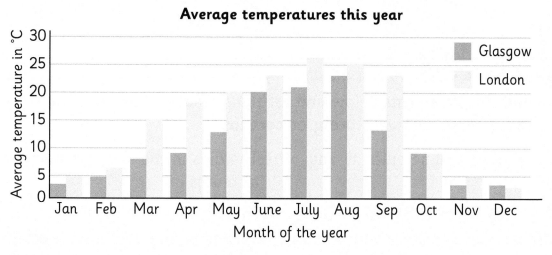

Average temperatures this year

Which of these statements is appropriate for this inquiry?
a) I found that Glasgow was always colder than London this year.
b) I found that London was twice as warm as Glasgow in March.
c) I wonder if it is warmer in London than Glasgow every year.
d) I think it rains a lot in Glasgow.

CHALLENGE!

Look at the data from a statistical inquiry you have carried out.
Write three statements based on your results.

1) Prediction: I think ...

2) New inquiry: I wonder ...

3) Conclusion: I found ...

16 Ideas of chance and uncertainty

16.1 Investigating the possible outcomes of random events

We are learning to investigate the possible outcomes of random events.

There are a number of different things that occur as a result of chance.

Let's learn

When we talk about the **chance** or the **probability** of something happening we mean the **likelihood** of this event actually taking place.

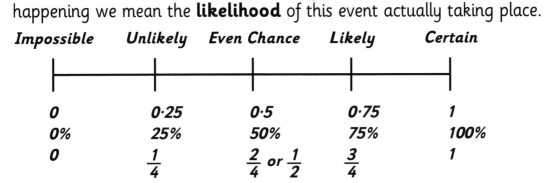

Impossible	Unlikely	Even Chance	Likely	Certain
0	0·25	0·5	0·75	1
0%	25%	50%	75%	100%
0	$\frac{1}{4}$	$\frac{2}{4}$ or $\frac{1}{2}$	$\frac{3}{4}$	1

There are some things that we know are certain:

there are 52 weeks in a year   there are 30 days in April

There are some things that we know are unlikely:

it will be warm in December   school will close early today

Let's practise

Using the words certain, likely, even chance, unlikely and impossible, answer the following questions:

1) Finlay places eight cards down numbered 1–8.

   If he randomly chooses one card how likely is it that he:

   a) Will choose the card with the 4 on it?
   b) Will choose a card with numbers 1–6 on it?
   c) Will choose a card with the number 9 on it?

2) Isla has two six-sided dice.

   If she rolls the dice, how likely is it that she:

   a) Rolls a total of 12?
   b) Rolls a total of between 2 and 8?
   c) Rolls a total of 13?

3) Parker was asked to choose a number from 1–9.

   What is the chance he chooses:

   a) The number 11?
   b) An odd number?
   c) An even number?

CHALLENGE!

Using the terms 0, 25%, 50%, 75% and 100%, work with a partner to investigate the following (*giving details of your choice of answer*):

- What is the probability that everyone in your class can swim?
- What is the probability that you will have the same size shoe as two other people in your class?
- What are the chances of all pupils in your class being absent on the same day?
- What are the chances that it will rain at playtime?

Answers

1 Estimation and rounding

1.1 Rounding whole numbers to the nearest 10, 100 or 1000 (p.2)

Before we start

Answers will vary.

Smallest number is 01234

1) a) 600 and 700. Rounded, 600.

   b) 7800 and 7900. Rounded, 7900.

2) a) Yes   b) Yes   c) No   d) Yes

3)

Month	Visitors	Nearest 10	Nearest 100	Nearest 1000
March	4568	4570	4600	5000
April	5429	5430	5400	5000
May	7347	7350	7300	7000

Challenge

Answers will vary.

1.2 Rounding decimal fractions to two places (p.4)

Before we start

5 represents five tenths and 6 is six hundredths

One tenth more is 2·66

One hundredth more is 2·57

Questions

1) a) 7   b) 0·5   c) 1·9 and 2

2)

	Nearest tenth	Nearest whole number
5·56	5·6	6
19·26	19·3	19
17·01	17·0	17
206·69	206·7	207
1·48	1·5	1

3) 1·31 rounded to the nearest tenth is 1·4   **false**

   85·55 rounded to the nearest tenth is 85·5   **false**

   10·04 rounded to the nearest tenth is 10·4   **false**

   3·71 rounded to the nearest whole number is 4   **true**

   19·78 rounded to the nearest whole number is 19·8   **false**

Challenge

We could round to the nearest pound to quickly estimate if she has enough money.

£1 + £4 + £1 + £3 = £9, so we could estimate that yes, £10 is enough to buy all these items.

1.3 Using rounding to estimate the accuracy of a calculation (p.6)

Before we start

24570   24600   25000

Questions

1) The amounts on the bill are approximately £14 + £20 + £20 + £14 = £68, so the bill must be incorrect.

2) a) 2300 + 5700 = 8000 so answer is not reasonable.

   b) 1700 − 300 = 1400 so answer is reasonable.

   c) 60 + 30 + 140 = 230 so answer is not reasonable.

   d) 240 + 540 = 780 so answer is not reasonable.

   e) 7900 − 3400 = 4500 so answer is reasonable.

   f) 9900 − 3600 = 6300 so answer is not reasonable.

3)

	Estimate	Higher or lower?	Correct answer	Check with estimate
3829 + 2990	6800	The answer will be higher.	6819	Yes, the answer was higher than the estimate.
3541 − 772	2700	The answer will be higher.	2769	Yes, the answer was higher than the estimate.
5403 + 4773	10200	The answer will be lower.	10176	Yes, the answer was lower than the estimate.
15432 − 8762	6600	The answer will be higher.	6670	Yes, the answer was higher than the estimate.

Challenge

Answers will vary.

2 Number – order and place value

2.1 Reading and writing whole numbers (p.8)

Before we start

Finlay is incorrect because he has added two extra zeros (he has written 9000 and 10 rather than 9010). Amman is incorrect because he has written nine thousand, one hundred. The boys should have written 9010.

Questions

1) a) twenty-four thousand, five hundred and eighty-six

   b) seventy-two thousand, nine hundred and ninety-eight

   c) thirty-one thousand, eight hundred

   d) ten thousand, three hundred and twelve

   e) fifty-five thousand, five hundred and five

   f) sixty thousand and twenty

   g) eighty-nine thousand and thirty

   h) ninety thousand

2) There are 90 possibilities!

   42000, 42101, 42103, 42105, 42501, 42503, 42505, 42901, 42903, 42905, 46000, 46101, 46103, 46105, 46501, 46503, 46505, 46901, 46903, 46905, 47000, 47101, 47103, 47105, 47501, 47503, 47505, 47901, 47903, 47905, 82000, 82101, 82103, 82105, 82501, 82503, 82505, 82901, 82903, 82905, 86000, 86101, 86103, 86105, 86501, 86503, 86505, 86901, 86903, 86905, 87000, 87101, 87103, 87105, 87501, 87503, 87505, 87901, 87903, 87905, 92000, 92101, 92103, 92105, 92501, 92503, 92505, 92901, 92903, 92905, 96000, 96101, 96103, 96105, 96501, 96503, 96505, 96901, 96903, 96905, 97000, 97101, 97103, 97105, 97501, 97503, 97505, 97901, 97903, 97905

Challenge

Nuria has used the 'spelling method'. She has read 300 and 56. The number has five digits and so is made up of tens of thousands, thousands, hundreds, tens and ones. She should have written thirty thousand and fifty-six.

2.2 Representing and describing whole numbers (p.10)

Before we start

a) and d) are correct.   b) is two thousand, two hundred.

c) is twenty thousand and twenty.

Questions

1) a) 5 thousands or 5000   b) 5 ones or 5

   c) 5 hundreds or 500   d) 5 tens or 50

   e) 50 thousands or 50 000   f) 5 thousands or 5000

2)

Number in numerals	Representation: place value arrow cards	Representation: place value counters
a) 13 666	10 000, 3000, 600, 60, 6	10 000, 1000, 1000, 1000, 100, 100, 100, 100, 100, 100, 10, 10, 10, 10, 10, 10, 1, 1, 1, 1, 1, 1
b) 44 202	40 000, 4000, 200, 2	10 000, 10 000, 10 000, 10 000, 1000, 1000, 1000, 1000, 100, 100, 1, 1
c) 70 011	70 000, 10, 1	10 000, 10 000, 10 000, 10 000, 10 000, 10 000, 10 000, 10, 1
d) 36 400	30 000, 6000, 400	10 000, 10 000, 10 000, 1000, 1000, 1000, 1000, 1000, 1000, 100, 100, 100, 100

3) a) 12 456   b) 65 421   c) 46 521   d) 24 156

   e) Both numbers have the digit 4 in the hundreds place. This means the 4 is worth 400.

Challenge

All of the following numbers are possibilities:

10 104	10 140	14 100	40 101	40 110
41 100	10 302	10 320	12 300	20 103
20 130	23 100	20 301	20 310	21 300
30 102	30 120	32 100		

2.3 Place value partitioning of whole numbers (p.12)

Before we start

a) Finlay has written 20 instead of 200. He should have written 3206 = 300 + 200 + 6.

b) Finlay has written 20 + 7 instead of 70 + 2. He should have written 8472 = 8000 + 400 + 70 + 2.

c) Finlay has written 600 instead of 60. He should have written 2063 = 2000 + 60 + 3.

d) Finlay has written 500 instead of 5000. He should have written 5005 = 5000 + 5.

Questions

1) a) 10 000 + 4000 + 500 + 90 + 4

   b) 50 000 + 8000 + 200 + 30 + 4

   c) 60 000 + 3000 + 400 + 80 + 7

   d) 20 000 + 7000 + 600 + 10 + 1

   e) 30 000 + 9000 + 900 + 20 + 5

   f) 40 000 + 100 + 10 + 3

   g) 70 000 + 8000 + 300 + 8

   h) 80 000 + 1000 + 70 + 5

2) a) 2 is worth 20 and 8 is worth 8000

   b) 7 is worth 700 and 3 is worth 30 000

   c) 9 is worth 9 and 4 is worth 4000

   d) 5 is worth 50 000 and 1 is worth 1000

3) a) 5 thousands, (0 hundreds), 8 tens and 3 ones

     4 thousands, 10 hundreds, 8 tens and 3 ones

     50 hundreds, 8 tens and 3 ones

   b) 4 thousands, 5 hundreds, 2 tens and 7 ones

     3 thousands, 15 hundreds, 2 tens and 7 ones

     45 hundreds, 2 tens and 7 ones

   c) 3 thousands, 7 hundreds, 1 ten and 5 ones

     2 thousands, 17 hundreds, 1 ten and 5 ones

     37 hundreds, 1 ten and 5 ones

   d) 2 thousands, 2 hundreds, 8 tens and 5 ones

     1 thousands, 12 hundreds, 8 tens and 5 ones

     22 hundreds, 8 tens and 5 ones

   e) 1 thousand, 2 hundreds, 1 ten and 0 ones

     12 hundreds, 1 ten and 0 ones

     121 tens and 0 ones

   f) 6 thousands, 0 hundreds, 0 tens and 0 ones

     5 thousands, 10 hundreds, 0 tens and 0 ones

     60 hundreds, 0 tens and 0 ones

4) a) 508 tens and 3 ones   b) 452 tens and 7 ones

   c) 371 tens and 4 ones   d) 228 tens and 5 ones

5) Two possible solutions are given for each question. Others are possible.

   a) 1560: 15 hundreds and 6 tens or 156 tens

   b) 1009: 10 hundreds and 9 ones or 100 tens and 9 ones

   c) 6816: 68 hundreds, 1 ten and 6 ones or 681 tens and 6 ones

   d) 9700: 97 hundreds or 970 tens

Challenge

Answers will vary.

2.4 Number sequences (p.16)

Before we start

a) The missing number is 10 each time. This works because the starting number and answer are both decreasing by 10 each time.

b) The missing number is decreasing by 100 each time. This works because the starting number is increasing by 100 each time but the result stays the same.

2690 − ☐10 = 2680	1499 + ☐500 = 1999
2680 − ☐10 = 2670	1599 + ☐400 = 1999
2670 − ☐10 = 2660	1699 + ☐300 = 1999

Questions

1) a) <u>51 450</u>, 51 451, 51 452, 51 453, 51 454

   b) 30 897, 30 898, 30 899, <u>30 900</u>, 30 901

   c) 76 996, 76 997, 76 998, 76 999, <u>77 000</u>

   d) 29 999, <u>30 000</u>, 30 001, 30 002, 30 003

2) The number is 59 890

   59 900, 59 990, 60 890, 69 890

   59 880, 59 790, 58 890, 49 890

3) a) 19 000   b) 28 087   c) 90 004   d) 77 777

   e) 12 590   f) 90 900   g) 29 000   h) 5000

Challenge

First row: 10 810, 10 820, 10 850, 10 880

Second row: 10 900, 10 940, 10 960

Third row: 11 000, 11 010, 11 020, 11 030, 11 060, 11 070, 11 090

2.5 Comparing and ordering whole numbers (p.18)

Before we start

679, 769, 967, 1679, 1769, 6719, 7617, 9671

9671, 7617, 6719, 1769, 1679, 967, 769, 679

Questions

1) a) 23 290 < 29 320   b) 10 389 > 10 144

    c) 34 277 < 73 472   d) 66 137 < 67 613

    e) 58 200 > 58 020   f) 83 038 > 83 030

2) a) Amsterdam (661 km), Barcelona (1666 km), Rome (1930 km), New York (5259 km), Delhi (6840 km), Singapore (10 930 km), Sydney (16 880 km), Auckland (17 905 km)

    b) Singapore is 9000 km further from Edinburgh than Rome.

Challenge

Nine numbers are possible: 49 645, 49 815, 49 239, 30 645, 30 815, 30 239, 78 645, 78 815, 78 239

Many answers are possible. Accept any statement in which the < or > has been used to correctly compare two of the above numbers, for example 49 815 < 78 815 or 30 815 > 30 239.

2.6 Identifying and placing positive and negative numbers (p.20)

Before we start

Monday

Day: *eleven degrees Celsius* or 11 degrees above freezing

Night: *four degrees Celsius* or 4 degrees above freezing

Tuesday

Day: *seven degrees Celsius* or 7 degrees above freezing

Night: *zero degrees Celsius* or freezing point

Wednesday

Day: *five degrees Celsius* or 5 degrees above freezing

Night: *minus one degree Celsius* or 1 degree below freezing

Thursday

Day: *three degrees Celsius* or 3 degrees above freezing

Night: *minus five degrees Celsius* or 5 degrees below freezing

Friday

Day: *one degree Celsius* or 1 degree above freezing

Night: *minus four degrees Celsius* or 4 degrees below freezing

Questions

1) a) −17   b) −13   c) −9   d) −3

    e) 0   f) 4   g) 11   h) 18

2) a) 42°C Forty-two degrees Celsius.

    b) −16°C Minus sixteen degrees Celsius.

    c) 6°C Six degrees Celsius.

    d) −4°C Minus four degrees Celsius.

    e) 30°C Thirty degrees Celsius.

3) a) −16°C, −8°C, −3°C, 0°C, 6°C, 10°C

    b) −14°C, −7°C, −5°C, 2°C, 8°C, 14°C

    c) −49°C, −28°C, −15°C, −4°C, −1°C, 0°C

Challenge

Mr Wang will be on Level −4 or four floors below ground level. The Burj Khalifa in Dubai is the tallest skyscraper in the world. It has 163 floors, only one of which is below ground level.

2.7 Reading and writing decimal fractions (p.22)

Before we start

a) ▢ is the odd one out because it shows five tenths. All of the others mean seven tenths.

b) $1\frac{4}{10}$ is the odd one out because all of the others mean one and six tenths.

222

Questions

1) a) We write 0·37 and we say *zero point three seven*.

    b) We write 1·08 and we say *one point zero eight*.

    c) We write 2·14 and we say *two point one four*.

    d) We write 0·62 and we say *zero point six two*.

    e) We write 1·99 and we say *one point nine nine*.

2) a) Eight point two six, 8.26 and $8\frac{26}{100}$

    b) Seven point three five, 7.35 and $7\frac{35}{100}$

    c) Two point zero two, 2.02 and $2\frac{2}{100}$

    d) Fifteen point nine nine, 15.99 and $15\frac{99}{100}$

Challenge

Nuria is incorrect because she thinks it says *eighty-two ninths*. She should have read this decimal fraction as *eight and ninety-two hundredths*. Finlay is also incorrect because you must read each digit after the decimal point separately. The correct way to read this decimal fraction is *eight point nine two*, not *eight point ninety two*.

2.8 Representing and describing decimal fractions (p.24)

Before we start

Both models show 2.4 because in each case two whole units and $\frac{4}{10}$ of a unit are shaded.

Questions

1) a) Each grid is divided into one hundred equal parts (hundredths). In each model one whole grid and $\frac{5}{100}$ of a grid are shaded.

    b) $\frac{95}{100}$ or 0.95 is unshaded.

2) Check decimal fractions are accurate on squared paper.

3)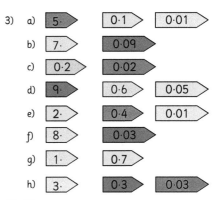

    a) 5·   0·1   0·01

    b) 7·   0·09

    c) 0·2   0·02

    d) 9·   0·6   0·05

    e) 2·   0·4   0·01

    f) 8·   0·03

    g) 1·   0·7

    h) 3·   0·3   0·03

Challenge

1) You would need to shade 80 hundredths.

2) You would need to shade 60 hundredths.

2.9 Comparing and ordering decimal fractions (p.26)

Before we start

The missing numbers are: 0·4, 1·1, 1·5 and 1·9. Various answers are possible, for example:

0·4 It has no ones and four tenths. It is the number between zero point three and zero point five on the decimal tenth number line, etc.

Questions

1) a) The arrows are pointing to: 3·72, 3·76, 3·8(0), 3·84, 3·89, 3·94

    b) The arrows are pointing to: 10·01, 10·05, 10·09, 10·12, 10·2(0), 10·22

2) a) true   b) false   c) false   d) false

    e) false   f) true   g) true   h) true

3) b) $15 \cdot 51 < 51 \cdot 15$ c) $9 > 7 \cdot 24$
   d) $18 = 18 \cdot 00$ e) $3 \cdot 31 > 3 \cdot 13$

4) a) The first missing whole number could be **0, 1** or **2**. The second missing whole number could be **4, 5, 6, 7** or **8**. The final missing whole number must be **9**.

   b) The first missing number must be **11**. The next two consecutive missing whole numbers could be **10, 9, 8, 7, 6, 5, 4** or **3**. The second number of this pair must be smaller. The final missing whole number could be **1** or **0**.

Challenge
$1 \cdot 29, 1 \cdot 9, 1 \cdot 92, 9, 9 \cdot 21, 9 \cdot 99, 19$

3 Number – addition and subtraction

3.1 Mental addition and subtraction (p.28)

Before we start
Note: Other mental strategies are possible

a) 384 The most efficient strategy is to add 50 then add 1.
b) 560 The most efficient strategy is to add 14 and 46 first.
c) 290 The most efficient strategy is to add 22 and 68 first.
d) 617 The most efficient strategy is to subtract 80 then add 1 back on.
e) 335 The most efficient strategy is to subtract 30 then add 2 back on.
f) 148 The most efficient strategy is to subtract 50 then add 1 back on.

Questions
1) a) 485 b) 562 c) 806 d) 782 e) 564
   f) 1212 g) 992 h) 1696 i) 1521 j) 1215
   The secret word is STRATEGIES.

2) a) 433 b) 405 c) 220 d) 1627 e) 2146
   f) 4445 g) 850 h) 2317 i) 1797

3) a) False. The numbers to the right have both been increased by 1 which increases the answer by 2.
   b) True. Addition is commutative which means the order does not matter.
   c) True, the numbers on the right have both been increased by 1 making the difference the same.
   d) False. One number has been changed.

Challenge
One has been added to the first number but 11 has been subtracted from the second number.
5 tens and 2 ones should be 6 tens and 4 ones making the equation $2741 - 1263 = 2742 - 1264$

3.2 Adding and subtracting a string of numbers (p.30)

Before we start
365 add 270 makes 635 **not** 5135. The correct answer is 635.
526 add 174 makes 700 **not** 690. The correct answer is therefore 125 ($825 - 700$).

Questions
1) a) 1868 b) 1901 c) 4851 d) 5000 e) 9099
   f) 8000 g) 9269 h) 6040 i) 8588 j) 9672

2) a) 4005 b) 7218 c) 6021 d) 5760 e) 2200
   f) 4488 g) 2000 h) 2001 i) 6200 j) 3100

3) a) 2740 b) 277

Challenge
1**00**6 + 3200 + 1240 = 5446   4625 − 1065 − **200** = 3000

3.3 Adding and subtracting multiples of 10, 100 and 1000 (p.32)

Before we start
a) $8214 + 1\underline{\mathbf{302}} = 9516$
b) $2730 + \underline{\mathbf{7}}000 = 9\underline{\mathbf{730}}$
c) $5049 - \underline{\mathbf{4046}} = 1003$
d) $\underline{\mathbf{7777}} - 3121 = 4656$

Questions
1) a) 31 922 b) 51 580 c) 72 103
   d) 41 546 e) 75 439 f) 107 234
   g) 5140 h) 64 198 i) 18 255

2) a) 5160 b) 36 217 c) 58 621
   d) 15 599 e) 60 316 f) 23 113
   g) 89 991 h) 95 760 i) 143 895

3) a) 3000 b) 1600 c) 66 900
   d) 55 700 e) 4000 f) 62 000

Challenge
43 500

3.4 Using place value partitioning (p.34)

Before we start
a) Missing number 120. Total 1332
b) Missing number 1000. Total 1161
c) Missing number 120 and 17. Total 837

Questions
1) a) 9986 b) 12 066 c) 14 338
   d) 26 296 e) 40 925 f) 26 068
   g) 74 844 h) 77 890 i) 93 206

2) a) 24 225 b) 16 237 c) 33 352
   d) 51 114 e) 17 272 f) 14 311
   g) 20 715 h) 10 256 i) 10 212

3) a) $7000 + 100 + 10 + 7 = 7117$
   b) $8000 + 400 + 590 + 8 = 8998$
   c) $5000 + 200 + 130 + 5 = 5335$
   d) $2000 + 320 + 340 + 2 = 2662$

Challenge
a) The calculation 8531 + 7420 gives the largest total (15 951)
b) The calculation 8731 + 5420 gives the second largest total (14 151)

3.5 Adding four-digit numbers using standard algorithms (p.36)

Before we start
1 thousand, 9 hundreds, 7 tens, 5 ones
19 hundreds, 7 tens, 5 ones
19 hundreds and 75 ones
1 thousand, 97 tens, 5 ones
1 thousand, 9 hundreds, 75 ones
1 thousand, 975 ones

Questions
1) a) 9193 b) 7821 c) 7740
   d) 6521 e) 10 761 f) 10 122
   g) 15 125 h) 12 876 i) 14 363

2) a) 8131 b) 13 554 c) 15 251
   d) 16 417 e) 14 488 f) 10 960

3) a) 5290 b) 11 145 c) 13 989 d) 5683

4) Working from the bottom up and from left to right the missing numbers are:

Fourth row:	3781	3777	4087	3858
Third row:	7558	7864	7945	
Second row:	15 422	15 809		
Top row:	31 231			

Challenge

Calculations completed with bold digits inserted correctly.

a)   4 7 8 6
    + 2 **9** 3 1
    ‾‾‾‾‾‾‾
      7 7 1 7

b)   3 7 5 5
    + 2 6 8 **9**
    ‾‾‾‾‾‾‾
      6 4 4 4

c)   **7 4 8 8**
    + 4 1 7 4
    ‾‾‾‾‾‾‾
    1 1 6 6 2

d)   5 8 2 **8**
    + 7 2 9 3
    ‾‾‾‾‾‾‾
    1 3 1 **2** 1

3.6 Column subtraction: three-digit numbers (p.38)

Before we start

Amman should have said:

If I start on 8 and take 10 jumps backwards I will be on −2.

If I start on 3 and take 7 jumps backwards I will be on −4.

Isla should have said:

If I start at 50 and count back 80 I will be on −30.

If I start at 10 and count back 70 I will be on −60.

Questions

1)   a) 509   b) 428   c) 181   d) 134
     e) 581   f) 227   g) 309   h) 174

2)   a) 294   b) 257   c) 74   d) 649
     e) 76   f) 389   g) 349   h) 439
     i) 195   j) 415   k) 86   l) 385

Challenge

Finlay has added 20 where he should have subtracted 20. 30 minus 50 equals minus 20. He should have written:

```
   6 3 5
 − 2 5 4
 ‾‾‾‾‾‾
       1
    − 2 0
   4 0 0
 ‾‾‾‾‾‾
   3 8 1
```

3.7 Column subtraction: four-digit numbers (p.40)

Before we start

a)   −100
b)   −500
c)   −600
d)   −700
e)   −700
f)   −300

Questions

1)   a) 3215   b) 4410   c) 3264
     d) 3994   e) 3474   f) 7521
     g) 2424   h) 2362   i) 1145

2)   a) 4551   b) 2818   c) 2138
     d) 5762   e) 1919   f) 4935
     g) 1341   h) 2777   i) 1697
     j) 3977   k) 1398   l) 586

Challenge

a)   2879   b) 859   c) 5748
d)   1756   e) 2352   f) 1558

With the exception of the thousands digits all other subtractions are negative numbers. For example, e.g. 2 − 3 = −1

60 − 80 = −20   300 − 400 = −100

3.8 Subtracting three-digit numbers using standard algorithms (p.42)

Before we start

2 boxes of 100, 3 packs of 10 and **4** single pencils.

1 box of 100, **13** packs of 10 and **4** single pencils.

2 boxes of 100, 2 packs of 10 and **14** single pencils.

Questions

1)   a) 542   b) 342   c) 223   d) 92
     e) 234   f) 221   g) 142   h) 394

2)   a) 278   b) 389   c) 139   d) 468
     e) 246   f) 348   g) 477   h) 356

3)   a) 179   b) 156   c) 763   d) 57
     e) 197   f) 567   g) 325   h) 67
     i) 645   j) 208   k) 246   l) 54

4)   CALCULATIONS

Challenge

The missing hundreds digit can only be 3.

The missing tens digits could be: 1 and 1, 2 and 0 or 0 and 2.

3.9 Subtracting four-digit numbers using standard algorithms (p.45)

Before we start

6 thousands, 13 hundreds, 6 tens

7 thousands, 36 tens

7 thousands, 3 hundreds, 60 ones

7 thousands, 3 hundreds, 6 tens

Questions (Standard)

1)   a) 542   b) 342   c) 223   d) 92
     e) 234   f) 221   g) 142   h) 394

2)   a) 278   b) 389   c) 139   d) 468
     e) 246   f) 348   g) 477   h) 356

3)   a) 179   b) 156   c) 763   d) 57
     e) 197   f) 567   g) 325   h) 67
     i) 645   j) 208   k) 246   l) 54

4)   Secret word is CALCULATIONS

Questions (Algorithm)

1)   a) 4095   b) 918   c) 1961
     d) 845   e) 6524   f) 2592
     g) 6660   h) 7451   i) 7829

2)   a) 3396   b) 5918   c) 675
     d) 3463   e) 4125   f) 3886
     g) 1667   h) 778   i) 1487

3)   a) 3059   b) 2575   c) 3862   d) 2376

Challenge

5762

3.10 Mental and written strategies (p.48)

Before we start

a)   6246   b) 17

Each answer should be accompanied by a reasonable justification for choice of strategy.

Questions

1)   Working from the bottom up and from left to right the missing numbers are:

Fourth row:	991	1217	1027	1108
Third row:	2208	2244	2135	
Second row:	4452	4379		
Top row:	8831			

2)   Answers will vary.

Challenge

a)   3395   b) 4820 and 3455   c) 2840, 4830 and 1435

3.11 Representing and solving word problems (p.50)

Before we start

The calculation 4785 + 5162 = 9947 correctly represented as a bar model, empty number line and algorithm.

Questions

1)   Each calculation correctly represented on a Think Board as a bar model, empty number line and algorithm.

a)   1924 − 1876 = 48 years

b)   1642 − 1593 = 49 years

c)   1712 + 160 = 1872, so the petrol engine was invented in 1872.

2) a) The difference between 1872 and 1887 is 15 years, therefore the first petrol-powered car was invented **15 years** after the petrol engine.

   b) 1903 – 1887 = 16 so the petrol-powered car was invented first, 16 years before the aeroplane.

   c) 2019 – 1879 = 140 years (note: for every year after 2019, add one year to the answer)

Challenge

Isla is incorrect. She has calculated 2 – 1 rather than 1 – 2. The correct answer is 259.

3.12 Solving multi-step word problems (p.52)

Before we start

848 pencils

Questions

1) 2000 tickets   2) £1112

3) No. Amman's brother has £57 left. Two seat covers would cost him £58.

4) £2865

5) 2270 tickets were sold on Wednesday.

Challenge

Answers will vary. The problem created should involve at least two steps.

3.13 Adding decimal fractions and whole numbers (p.54)

Before we start

a) 2·5, 2·6, 2·7, 2·8, 2·9, 3, 3·1, 3·2, 3·3

b) 4·5, 5·5, 6·5, 7, 8, 8·5, 9

Questions

1) a) 67·4   b) 46·3   c) 104·1   d) 510·6
   e) 182·7 f) 141·9 g) 87·5   h) 71·7

2) a) 19·1   b) 64·6   c) 47·2   d) 65·4
   e) 36·1   f) 72·5   g) 59·1   h) 34·1

3) a) 8·9   b) 14·8   c) 15·6   d) 57·1
   e) 36·7   f) 73·1   g) 27·6   h) 55·7

4) a) 12·7   b) 7·3   c) 8·2
   d) 38·1   e) 57·3   f) 58·5

Challenge

1000, 100, 10, 1
700, 30, 10, 0·3 + 0·7

3.14 Adding and subtracting decimal fractions (p.56)

Before we start

a) 6 + 0·9   5 + 1·9   4 + 2·9   3 + 3·9
   2 + 4·9   1 + 5·9   0 + 6·9

b) Multiple answers are possible. The ones must total 6 and the tenths must total 9. For example, 4·4 + 2·5 = 6·9.

Questions

1) a) 15·7   b) 12·2   c) 8·5   d) 21·8
   e) 17·6   f) 46·9   g) 8·6   h) 14·8
   i) 20·2   j) 47·6   k) 63·3   l) 14·4

2) a) 43 = 14·9 + 28.1   43 = 28·1 + 14·9
      43 – 14·9 = 28·1   43 – 28·1 = 14·9
   b) 71 = 50·5 + 20·5   71 = 20·5 + 50·5
      71 – 50·5 = 20·5   71 – 20·5 = 50·5
   c) 90 = 45·9 + 44·1   90 = 44·1 + 45·9
      90 – 44·1 = 45·9   90 – 45·9 = 44·1
   d) 100 = 69·4 + 30·6   100 = 30·6 + 69·4
      100 – 69·4 = 30·6   100 – 30·6 = 69·4

3) a) 21·4   b) 41·4   c) 25·8
   d) 52   e) 13·2   f) 16·6

Challenge

Any four decimal fractions that total 10, for example
5·1 + 1·3 + 2·2 + 1·4

4 Number – multiplication and division

4.1 Recalling multiplication and division facts for 6 (p.58)

Before we start

12 × 2 = 10 × 2 + 2 × 2 = 20 + 4 = 24

Questions

1) a) 2 × (3 × 2) = 12   b) 2 × (5 × 3) = 30
   c) 2 × (3 × 6) = 36   d) 2 × (3 × 8) = 48

2) a) 42 ÷ 6 = 7   b) 6 times 10 = 60
   c) five lots of 6 = 30   d) 18 divided by 6 = 3
   e) 1 × 6 = 6   f) 8 groups of 6 = 48
   g) 9 × 6 = 54   h) 36 split into 6 = 6
                               24 not used

3) a) 4   b) 54   c) 42   d) 2   e) 10
   f) 6   g) 3   h) 8   i) 5   j) 1

Challenge

Answers will vary.

4.2 Recalling multiplication and division facts for (p.60)

Before we start

32 = 4 × ? and I know 4 × 8 = 32

Questions

1) Ten 9s = 90
   Three groups of 9 = 27
   54 divided by 9 = 6
   8 × 9 = 72
   81 split into 9 = 9
   18 shared into 9 = 2
   Five 9s = 45
   4 × 9 = 36

2) Check that students have coloured in the table correctly, and identified the pattern formed.

3) Answers will vary.

Challenge

1 × 9 = 9   2 × 9 = 18   3 × 9 = 27
4 × 9 = 36   5 × 9 = 45
9 = 1 + 8 = 2 + 7 = 3 + 6 = 4 + 5
6 × 9 = 54   7 × 9 = 63   8 × 9 = 72
9 × 9 = 81   10 × 9 = 90

Add up to 9 again. If we try larger multiples, e.g. 14 × 9 = 126 still add up to 9.

87 × 9 = 783 which add up to 18.

They will always add up to a multiple of 9.

4.3 Multiplying multiples of 10, 100 or 1000 (p.62)

Before we start

Isla is correct as dividing by 10 removes only one zero.

Questions

1) 40 × 4 = 160 using 4 × 4 = 16
   900 × 5 = 4500 using 9 × 5 = 45
   800 × 3 = 2400 using 8 × 3 = 24
   70 × 2 = 140 using 7 × 2 = 14
   500 × 5 = 2500 using 5 × 5 = 25
   60 × 3 = 180 using 6 × 3 = 18
   800 × 4 = 3200 using 8 × 4 = 32
   90 × 2 = 180 using 9 × 2 = 18
   60 × 5 = 300 using 6 × 5 = 30

2) Answers such as 90 × 4 = 360 or 9 × 400 = 3600

3) a) 450 tickets   b) 1500 tickets   c) 10 000 tickets

Challenge

With two zeros it could be 600 × 6 or 60 × 60 or 6 × 600
With three zeros it could be 6000 × 6, 600 × 60, 60 × 600,
6 × 6000

4.4 Dividing by multiples of 10, 100 or 1000 (p.64)

Before we start

11 × 100 = 1100, 12 × 10 = 120 and 26 × 1 = 26
1100 + 120 + 26 = 1246 so £1246

Questions

1) a) 24 ÷ 6 and 40   b) 45 ÷ 5 and 90
   c) 27 ÷ 3 and 90   d) 24 ÷ 4 and 60

2) 300 ÷ 6 = 50   150 ÷ 3 = 50   2400 ÷ 6 = 400
   1200 ÷ 2 = 600   300 ÷ 3 = 100   1500 ÷ 5 = 300

3) 540 ÷ 6 = 54 tens ÷ 6
            = 9 tens
            = 90

Challenge

200 ÷ 50 = 4
a) 120 ÷ 60 = 2   b) 150 ÷ 30 = 5   c) 210 ÷ 30 = 7

4.5 Solving division problems (p.66)

Before we start

6 × 6 = (2 × 3) × 6 = 2 × (3 × 6) = 2 × 18 = 36

Questions

1) a) 5 × 2 = 10   so   10 ÷ 2 = 5
   b) 5 × 6 = 30   so   30 ÷ 6 = 5
   c) 5 × 9 = 45   so   45 ÷ 9 = 5

2) a) 8 × 6 = 48   so   48 ÷ 8 = 6
   b) 4 × 12 = 48   so   48 ÷ 4 = 12
   c) 3 × 16 = 48   so   48 ÷ 3 = 16

3) a) 40 ÷ 2 = 20   b) 21 ÷ 3 = 7   c) 40 ÷ 5 = 8

Challenge

Many answers are possible such as 2 × 60 = 120 so
120 ÷ 2 = 60 or 120 ÷ 60 = 2

4.6 Solving multiplication problems (p.68)

Before we start

32 × 10 = 320 so £320

Questions

1) a) 600 + 150 + 18 = 768
   b) 2000 + 450 + 40 = 2490
   c) 2000 + 80 + 16 = 2096
   d) 1800 + 420 + 30 = 2250
   e) 2400 + 180 + 12 = 2592
   f) 4800 + 560 + 32 = 5392

2) a) 634 × 4 = (600 + 30 + 4) × 9 = (600 × 4)
      + (30 × 4) + (4 × 4) = 2400 + 120 + 16 = 2536
   b) 752 × 6 = (700 + 50 + 2) × 6 = (700 × 6)
      + (50 × 6) + (2 × 6) = 4200 + 300 + 12 = 4512
   c) 372 × 5 = (300 + 70 + 2) × 5 = (300 × 5)
      + (70 × 5) + (2 × 5) = 1500 + 350 + 10 = 1860
   d) 284 × 6 = (200 + 80 + 4) × 6 = (200 × 6)
      + (80 × 6) + (4 × 6) = 1200 + 480 + 24 = 1704
   e) 378 × 4 = (300 + 70 + 8) × 4 = (300 × 4)
      + (70 × 4) + (8 × 4) = 1200 + 280 + 32 = 1512

3) a) 165 × 9 = (100 × 9) + (60 × 9) + (5 × 9)
              = 900 + 540 + 45 = 1485
   b) 352 × 9 = (300 × 9) + (50 × 9) + (2 × 9)
              = 2700 + 450 + 18 = 3168

c) 541 × 9 = (500 × 9) + (40 × 9) + (1 × 9)
           = 4500 + 360 + 9 = 4869

Challenge

There are six possibilities:

245 × 6	254 × 6	246 × 5
264 × 5	256 × 4	265 × 4

4.7 Using partitioning to solve division problems (p.70)

Before we start

Nuria is correct as 90 × 5 = 9 tens × 5 = 45 tens = 450 and
9 × 50 = 9 × 5 tens = 45 tens = 450

Questions

1)

Problem	Partition the larger number	Enter the values in the grid		Answer
64 ÷ 4	64 = 40 + 24	40	24	10 + 6 = 16
		4 10	6	
96 ÷ 8	96 = 80 + 16	80	16	10 + 2 = 12
		8 10	2	
84 ÷ 6	84 = 60 + 24	60	24	10 + 4 = 14
		6 10	4	

2)

Problem	Partition	Grid		Answer
78 ÷ 3	78 = 60 + 18	60	18	20 + 6 = 26
		3 20	6	
84 ÷ 7	84 = 70 + 14	70	14	10 + 2 = 12
		7 10	2	
75 ÷ 5	75 = 50 + 25	50	25	10 + 5 = 15
		5 10	5	

3) a) 126 ÷ 6 = 21   b) 126 ÷ 9 = 14   c) 126 ÷ 3 = 42
   Working should show recording of partitioning,
   e.g. 126 ÷ 6 = (120 ÷ 6) + (6 ÷ 6)
                = 20 + 1
                = 21

Challenge

Answers will vary.

4.8 Using rounding and compensating to solve multiplication problems (p.73)

Before we start

4 × 14 = 56 which is 60 rounded to the nearest 10

Questions

1) a) 5 × 10 = 50 50 − 5 = 45 so 5 × 9 = 45
   b) 7 × 10 = 70 70 − 7 = 63 so 7 × 9 = 63
   c) 4 × 10 = 40 40 − 4 = 36 so 4 × 9 = 36
   d) 6 × 10 = 60 60 − 6 = 54 so 6 × 9 = 54

2) a) 174   b) 186   c) 168

3) a) £96   b) £240   c) £288   d) £432

Challenge

Answers could be 21 × 5, 21 × 6, 21 × 7, 21 × 8, 21 × 9

4.9 Solving problems involving addition, subtraction, multiplication and division (p.76)

Before we start

Answers will vary but should include methods to get
16 × 5 = 80

Questions

1) a) $4 \times 8 = 32$ (Isla)
    $6 \times 8 = 48$ (Finlay)
   b) $48 - 32 = 16$
    Finlay picked
    16 more punnets.

2) a) £8 b) £15 c) £28
   d) £18 e) £20 f) £89

3) a) £29 b) £34
   c) The first family would be £1 more for a family ticket. The second family would be £4 less for a family ticket.

Challenge

Answers will vary.

4.10 Multiplying decimal fractions (p.78)

Before we start

b) $40 \times 5 = 200$ c) $4 \times 500 = 2000$
d) $4000 \times 5 = 20\,000$

Questions

1) a) $6{\cdot}0$ b) $8{\cdot}0$ c) $9{\cdot}0$

2) a) 57 b) 620 c) 8800
   d) 1220 e) 10 f) 1000

3) a) 320 cm b) 1790 cm c) 80 cm d) 1010 cm

Challenge

Kim started with 400
We can work it out by going backwards through the problem
$? \div 10 = 2{\cdot}4$ so $? = 2{\cdot}4 \times 10 = 24$
$? + 20 = 24$ so $? = 24 - 20 = 4$
$? \div 100 = 4$ so $? = 4 \times 100 = 400$

4.11 Dividing whole numbers by 10, 100 and 1000 (p.80)

Nuria is right as dividing by 10 means taking off one zero

Questions

1) a) $1{\cdot}7$ b) $2{\cdot}1$ c) $4{\cdot}8$ d) $6{\cdot}6$
2) a) $46 \div 10 = 4.6$ b) $83 \div 10 = 8{\cdot}3$
   c) $427 \div 10 = 42{\cdot}7$ d) $903 \div 10 = 90{\cdot}3$
   e) $80 \div 100 = 0{\cdot}8$ f) $660 \div 100 = 6{\cdot}6$
   g) $8100 \div 1000 = 8{\cdot}1$ h) $900 \div 1000 = 0{\cdot}9$

3) a) $0{\cdot}2$ litres b) $1{\cdot}8$ litres c) 3 litres d) $5{\cdot}9$ litres

Challenge

Isla started with the number $0{\cdot}8$.

4.12 Solving division problems with remainders (p.82)

Before we start

Order is 0.18 $\frac{1}{2}$ 4.19   4.2   4.31 $5\frac{1}{2}$

Questions

1) a) six teams ($32 \div 5 = 6$ r 2)
   b) It is most useful to write the remainder as whole numbers to show how many teams and how many children are left over.

2) a) $6{\cdot}50$. It is easier to write the remainder as a decimal because the problem is about money.
   b) $6{\cdot}5$ km. It is easier to write the remainder as a decimal when the problem is about measuring distance.
   c) $6\frac{1}{2}$ biscuits. It is easier to write the remainder as a fraction as the biscuits could be halved to make it fair.
   d) 6 r 2. It is easier to write the remainder as a whole number to work out how many tables are needed.

3) a) $30 \div 12 = 2$ r 6 so 3 weeks altogether
   b) $59 \div 12 = 4$ r 11 so 5 weeks altogether
   c) $100 \div 12 = 8$ r 4 so 9 weeks altogether

Challenge

Answers will vary but should be sensible such as
e.g. $45 \div 10 = 4{\cdot}5$ when using a money problem
$48 \div 5 = 9$ r 3 when we have a whole number problem
$28 \div 8 = 3\frac{1}{2}$ when we have a sharing problem
$33 \div 4 = 8.25$ when using a money problem

4.13 Solving multiplication and division problems (p.84)

Before we start

$48 \div 12 = 4$ so four tins

Questions

1) a) $58 \times 6 = (60 \times 6) - (2 \times 6)$
    $= 360 - 12$
    $= 348$
   b) $702 \times 8 = (700 \times 8) + (2 \times 8)$
    $= 5600 + 16$
    $= 5616$
   c) $611 \times 4 = (600 \times 4) + (11 \times 4)$
    $= 2400 + 44$
    $= 2444$

2) Fast Track:
   $540 \div 9 = (54 \div 9) \times 10 = 60$ (NO)
   Safe Load:
   $220 \div 4 = $ Half of 220 then halved again $= 55$ (YES)
   Quick Star:
   $275 \div 5 = (250 \div 5) + (25 \div 5) = 50 + 5 = 55$ (YES)

3) Students' answers will vary.

Challenge

$102 \div 6 = 17$

5 Multiples, factors and primes

5.1 Identifying all factors of a number (p.87)

Before we start

$4 \times 6 = 24$ so Nuria has £24

Questions

1) a) 1, 2, 3, 6 b) 1, 2, 5, 10 c) 1, 5, 25

2) Only number 9 should be left.

3) a) 2 b) 4 c) 5 d) 7

Challenge

60

5.2 Identifying multiples of numbers (p.89)

Before we start

20 and 70 we might write for example $2 \times 10 = 20$, $70 \div 10 = 7$

Questions

1) 23   51   65   114   54

2) a) 3, 6, 9, 12, 15, 18, 21, 24, 27, 30
   b) 6, 12, 18, 24, 30, 36, 42, 48, 54, 60
   c) 9, 18, 27, 36, 45, 54, 63, 72, 81, 90

3) 63 is a multiple of 9. **True**
   The fourth multiple of 2 is 12. **False**
   The multiples of 7 are all odd numbers. **False**
   40 is not a multiple of 10. **False**
   12 is a multiple of 3 and a multiple of 4. **True** **227**

The tenth multiple of 10 is 100.   **True**

Multiples of 8 are always even.   **True**

6 is a multiple of 12.   **False**

4)   Check colouring is accurate. Answers may vary.

Challenge

The lowest common multiple of 2 and 3 is 6.

The lowest common multiple of 3 and 5 is 15.

The lowest common multiple of 2 and 5 is 10.

6 Fractions, decimal fractions and percentages

6.1 Identifying equivalent fractions (p.91)

Before we start

Answers will vary but could be $\frac{2}{6}, \frac{3}{9}, \frac{10}{30}$

Questions

1) a) No: one third can be converted into three ninths and four twelfths.

    b) Yes: two quarters can be converted into one half, which can be converted into five tenths.

    c) Yes: one fifth can be converted into two tenths.

    d) Yes: three fifths can be converted into six tenths.

    e) No: one sixth can be converted into two twelfths.

    f) No: three eighths can be converted into six sixteenths.

2) a) Yes: one half can be converted into 50 hundredths.

    b) Yes: three quarters can be converted into 75 hundredths.

    c) No.

    d) Yes: one fifth can be converted into 20 hundredths.

    e) Yes: one tenth can be converted into 10 hundredths.

    f) No.

Challenge

Yes: Four eighths = one half = five tenths

Yes: Two quarters = one half = five tenths

No: One third

Yes: Four twentieths = two tenths

Yes: Three sixths = one half = five tenths

No: Three quarters

6.2 Calculating equivalent fractions (p.94)

Before we start

Answers will vary but could be $\frac{10}{12}, \frac{15}{18}, \frac{50}{60}$

Questions

1) b)

    c)

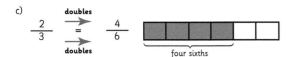

    d)

2) b) $\frac{1}{5} = \frac{3}{15}, \frac{4}{20}, \frac{5}{25}$ c) $\frac{2}{3} = \frac{6}{9}, \frac{8}{12}, \frac{10}{15}$

    d) $\frac{7}{10} = \frac{21}{30}, \frac{28}{40}, \frac{35}{50}$

Challenge

Answers will vary but might include

$\frac{1}{2} = \frac{2}{4} = \frac{3}{6} = \frac{4}{8} = \frac{10}{20}$ etc.

$\frac{3}{5} = \frac{2}{20} = \frac{9}{15} = \frac{12}{20} = \frac{30}{50}$ etc.

6.3 Comparing and ordering fractions (p.96)

Before we start

Order is $\frac{1}{10}$ $\frac{1}{5}$ $\frac{1}{4}$ $\frac{1}{3}$ $\frac{1}{2}$

Questions

1) a) seven tenths   b) nine twelfths   c) three tenths

2) $\frac{1}{2} = \frac{6}{12}$ so $\frac{5}{12}$ is less than $\frac{1}{2}$

    $\frac{1}{2} = \frac{10}{20}$ so $\frac{11}{20}$ is greater than $\frac{1}{2}$

    $\frac{1}{2} = \frac{4}{8}$ so $\frac{5}{8}$ is greater than $\frac{1}{2}$

    $\frac{1}{2} = \frac{25}{50}$ so $\frac{30}{50}$ is greater than $\frac{1}{2}$

    $\frac{1}{2} = \frac{50}{100}$ so $\frac{45}{100}$ is less than $\frac{1}{2}$

3) a)

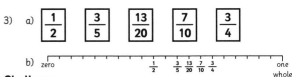

    b) zero $\frac{1}{2}$ $\frac{3}{5}$ $\frac{13}{20}$ $\frac{7}{10}$ $\frac{3}{4}$ one whole

Challenge

a) Amman ate three eighths; Finlay ate two eighths; Amman ate more cake.

b) Nuria drinks 10 twelfths; Isla drinks nine twelfths; Nuria drank more orange juice.

6.4 Decimal equivalents to tenths and hundredths (p.99)

Before we start

$\frac{3}{4} = \frac{75}{100}$ $\frac{4}{5} = \frac{8}{10} = \frac{80}{100}$ $\frac{6}{20} = \frac{3}{10} = \frac{30}{100}$

$\frac{2}{3}$ cannot be done $\frac{15}{20} = \frac{75}{100}$ $\frac{5}{8}$ cannot be done

Questions

1) a) 0·35

    b) 5 tenths + 4 hundredths = 0·54

    c) 7 tenths + 8 hundredths = 0·78

2) a) 32 hundredths = 0·32   b) 54 hundredths = 0·54

    c) 70 hundredths = 0·7(0)   d) 85 hundredths = 0·85

3) a) 2 whole and forty-five hundredths or 245 hundredths or 2·45

    b) 2 whole and seventy-four hundredths or 274 hundredths or 2·74

    c) 4 whole and ninety-nine hundredths or 299 hundredths or 2·99

Challenge

Nuria is right though some students may realise that thousandths can be used.

$\frac{1}{2} = 0·5$ $\frac{1}{3}$ cannot be done $\frac{1}{4} = 0·25$ $\frac{1}{5} = 0·2$

$\frac{1}{6}$ cannot be done $\frac{1}{8} = 0·125$ $\frac{1}{10} = 0·1$

6.5 Decimal equivalents to simple fractions (p.102)

Before we start

Answers will vary but might include

$\frac{1}{2} = \frac{2}{4} = \frac{3}{6} = \frac{4}{8} = \frac{10}{20}$ etc.

$\frac{1}{4} = \frac{2}{8} = \frac{3}{12} = \frac{4}{16} = \frac{10}{40}$ etc.

Questions

1) a) four fifths = eight tenths = 0·8
   b) one and a half = 1 and five tenths (or 15 tenths) = 1·5
   c) one quarter = twenty–five hundredths = 0·25

2) b) three quarters = seventy–five hundredths = 0·75
   c) three fifths = six tenths = 0·6
   d) two eighths = one quarter = twenty–five hundredths = 0·25

Challenge

$\frac{3}{6} = \frac{1}{2} = \frac{5}{10} = 0·5$ $\frac{6}{8} = \frac{3}{4} = \frac{75}{100} = 0·75$

$\frac{3}{12} = \frac{1}{4} = \frac{25}{100} = 0·25$ $\frac{14}{20} = \frac{7}{10} = 0·7$

$\frac{10}{50} = \frac{1}{5} = \frac{2}{10} = 0·2$ $\frac{15}{25} = \frac{3}{5} = \frac{6}{10} = 0·6$

6.6 Adding and subtracting fractions (p.105)

Before we start

Draw a larger bar for Amman.

Questions

1) $\frac{1}{3} + \frac{1}{3} = \frac{2}{3}$ $\frac{6}{7} + \frac{2}{7} = \frac{8}{7}$

   $\frac{1}{2} + \frac{2}{2} = \frac{3}{2}$ or $1\frac{1}{2}$ $\frac{7}{4} - \frac{3}{4} = \frac{4}{4}$ or 1

2) a) $\frac{3}{5}$ b) $\frac{4}{8}$ (or $\frac{1}{2}$)   c) $\frac{10}{10}$ (or 1)

   d) $\frac{2}{3}$ e) $\frac{6}{4}$ (or $1\frac{2}{4}$ or $1\frac{1}{2}$)

3) a) five sixths
   b) five tenths (or one half)
   c) twelve eighths (or one and a half)
   d) two quarters (or one half)

Challenge

$\frac{1}{3} + \frac{1}{6} = \frac{3}{6}$ (or $\frac{1}{2}$) $\frac{7}{8} - \frac{1}{4} = \frac{5}{8}$

$\frac{1}{2} + \frac{4}{12} = \frac{10}{12}$ (or $\frac{5}{6}$) $\frac{9}{10} - \frac{2}{5} = \frac{5}{10}$ (or $\frac{1}{2}$)

6.7 Calculating a fraction of a value (p.108)

Before we start

A bar divided into 3 so $42 \div 3 = 14$ in each so $\frac{2}{3}$ of 42 = 28

A bar divided into 4 so $96 \div 4 = 24$ in each so $\frac{3}{4}$ of 96 = 72

Questions

1) a) 69   b) 130   c) 335   d) 3045
2) Answers will vary.
3) a) 840 km   b) 2000 litres   c) 4131 steps

Challenge

a) Total population: 840   b) 1185; 810; 1056; 3750
c) Answers will vary.

6.8 Comparing numbers with two decimal places (p.111)

Before we start

Answers must be of the form 9·13 etc.

Questions

1) a) 0·34 is less than 0·42
   b) 1·12 is greater than 0·89
   c) 2·26 is greater than 2·17

2) a) Check drawing of number line - order should be 0·25, 0·29, 0·31, 0·34, 0·36, 0·37, 0·39, 0·40, 0·42, 0·45
   b) Answers will vary.
   c) Note that the smallest time is fastest.

   Jamaica   USA   GB   Canada   France
   South Africa   China   Australia

Challenge

Name	Score	Position
1Amman1	28·34	1st
James007	28·19	2nd
Finlay999	28·07	3rd
Nuria247	27·70	4th
Isla123	27·65	5th
F.I.N.N.0.8	27·53	6th

Answers will vary but they must score between 28·19 and 28·34

Difference first to last is $28·34 - 27·53 = 0·81$

6.9 Percentage (p.114)

Before we start

$\frac{3}{10}$ $\frac{42}{100}$ $\frac{76}{100}$

Questions

1) a) i) $\frac{90}{100} = 90\%$ ii) $\frac{25}{100} = 25\%$

   iii) $\frac{78}{100} = 78\%$ iv) $\frac{53}{100} = 53\%$

   b) i) 10%   ii) 75%   iii) 22%   iv) 47%

2) b) $\frac{13}{100}$ c) $\frac{72}{100}$ d) $\frac{85}{100}$

Challenge

25%; 20%; 10%; 4%

6.10 Converting fractions to percentages (p.116)

Before we start

$\frac{2}{10} = \frac{20}{100}$ $\frac{3}{5} = \frac{60}{100}$ $\frac{1}{4} = \frac{25}{100}$

Questions

1) a) 25%   b) 60%   c) 90%   d) 75%

2) a) $\frac{70}{100} = \frac{7}{10}$ b) $\frac{40}{100} = \frac{2}{5}$

   c) $\frac{25}{100} = \frac{1}{4}$ d) $\frac{75}{100} = \frac{3}{4}$

Challenge

a) $\frac{3}{5} = 60\%$ $\frac{1}{3} = 33.33\%$ $\frac{7}{10} = 70\%$

$\frac{5}{8} = 62.5\%$ $\frac{3}{6} = 50\%$

$\frac{4}{20} = 20\%$ $\frac{1}{7} = 14.29\%$ $\frac{12}{50} = 24\%$

$\frac{6}{8} = 75\%$ $\frac{1}{100} = 1\%$

6.11 Percentage calculation (p.119)

Before we start

$\frac{4}{10}$ of 20 is eight sweets

Questions

1) a) 30 boys; 30 girls   b) 10 boys; 40 girls
    c) 12 boys; 8 girls   d) 12 boys; 28 girls

2) a) £7   b) 56 sheep
    c) £10   d) 240 children

Challenge

a) 48 children can swim the backstroke

b) Answers will vary but must include 80% and 60%.

7 Money

7.1 Money problems using the four operations (p.122)

Before we start

a) £13·70   b) £4·30

Questions

1) a) £7·75   b) £5, £2, 50p, 20p, 5p

2) a) £15·50   b) No

3) a) £102·30   b) £17·70   c) £10, £5, £2, 50p, 20p

Challenge

a) £60 in total   b) £50·80   c) £4·00

7.2 Budgeting (p.124)

Before we start

a) 15 weeks   b) 1 week

Questions

1) a) Online   b) 25p

2) a) Diamond Drivers   b) £240   c) £24

Challenge

a) £6   b) £8·50

c) For six lessons she pays £30 + 6 × 0·50 for skates = £33

7.3 Profit and loss (p.126)

Before we start

£5 per chair

Questions

1) a) Loss   b) 40p

2) 67p or more

3) a) Loss of £10   b) £115

Challenge

a) 50p   b) 60p

c) 80p   d) 40 cups at £1 each/various

7.4 Discounts (p.128)

Before we start

He can get everything he needs but not everything he wants. He cannot get coloured pencils or gel pens.

Questions

1) a) £14   b) £45

2) a) £12   b) £108

3) £150

Challenge

a) £25   b) £45

7.5 Credit, debit and debt (p.130)

Before we start

Debt   a sum of money that is owed

Credit   a sum of money that is received or given

Debit   money that is taken out of a bank account

Questions

1) a) Security – so no-one can use your card.

   b) Answers will vary.   c) Answers will vary.

2) A debit card uses money that you have, but a credit card uses money borrowed from a bank.

3) Answers will vary.

Challenge

Answers will vary, including definitions of terms.

8 Time

8.1 Reading and writing 12-h and 24-h time (p.132)

Before we start

a) Eight minutes past four   b) Eight minutes past ten

Questions

1)

12-hour clock	24-hour clock
12·00 pm	**1200**
3·30 am	0330
5·21 pm	**1721**
8·14 am	0814
12·00 am	**0000**

2) a) $\frac{3\cdot14 \text{ pm}}{1514}$ b) $\frac{3\cdot55 \text{ am}}{0355}$ c) $\frac{10\cdot35 \text{ pm}}{2235}$

   d) $\frac{7\cdot45 \text{ am}}{0745}$ e) $\frac{12\cdot00 \text{ am}}{0000}$ f) $\frac{1\cdot57 \text{ pm}}{1357}$

3) b)   e)   a)   c)   d)

Challenge

Remember that the 5 can appear in the 2nd, 3rd or 4th place in the time. Answers will vary, including 170 or 504.

8.2 Converting units of time using fractions (p.134)

Before we start

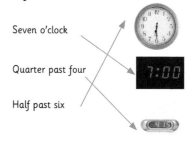

Seven o'clock

Quarter past four

Half past six

Questions

1) b) $\frac{1}{2}$ past eight

2) c) 75 minutes

3) c) 7·45 am

Challenge

45 minutes

8.3 Calculating time intervals using timetables (p.136)

Before we start

1) 12·45
2) 16·45
3) Sundays and public holidays

Questions

1) a) 14 mins   b) 16·23   2) 12·35

Challenge

a) 5 hrs 36 mins   b) 18 mins   c) 1 hr 36 mins

8.4 Measuring time (p.138)

Before we start

Answers will vary.

Questions

1) Answers will vary.
2) c) Stopwatch
3) e) Ruler
4) Answers will vary.

Challenge

Answers will vary.

8.5 Speed, distance and time calculations (p.140)

Before we start

160 miles

Questions

1) a) cm   b) miles   c) m
2) a) secs   b) hours
   c) secs or mins depending on the size of the playground
3) 125 miles
4) 87·5 miles

Challenge

67·5 miles

8.6 Time problems (p.142)

Before we start

a) false   b) true   c) false   d) true

Questions

1) 11 mins 20 secs   2) 250 mins
3) a) 30 mins   b) 24 mins   c) 20 mins

Challenge

a) 720 hrs   b) 43 200 mins   c) 2 592 000 secs

9 Measurement

9.1 Estimating and measuring length (p.144)

Before we start

Blue – 3 cm, Yellow – 7 cm, Green – 10 cm

Questions

1) a) 50 mm or 5 cm   b) 30 mm or 3 cm
   c) 60 mm or 6 cm   d) 80 mm or 8 cm
   e) 100 mm or 10 cm   f) 120 mm or 12 cm

2) Finn = 2·9 km; Ava = 1·6 km; Marissa = 2·1 km; Stuart = 3·7 km;
   Lauren = 4·2 km; Hamza = 3·0 km; Mark = 0·9 km

3) b) 2·4 m   c) 3·9 m   d) 6·7 m
   e) 2·0 m   f) 0·8 m

Challenge

Answers will vary.

9.2 Estimating and measuring mass (p.148)

Before we start

Answers will vary.

Questions

1) a) 800 g = 0·8 kg   b) 700 g = 0·7 kg
   c) 1300 g = 1·3 kg   d) 2200 g = 2·2 kg
2) Answers will vary.

Challenge

Answers will vary.

9.3 Estimating and measuring capacity (p.150)

Before we start

2500 ml and 3800 ml
Total is 6300 ml

Questions

1) b) 200 ml = 0·2 L   c) 900 ml = 0·9 L
   d) 1500 ml = 1·5 L   e) 1800 ml = 1·8 L
   f) 2400 ml = 2·4 L   g) 4300 ml = 4·3 L
   h) 4900 ml = 4·9 L

2) b) 0·4 L   c) 0·6 L   d) 0·8 L
   e) 1·2 L   f) 1·7 L

Challenge

Answers will vary.

9.4 Converting metric units (p.153)

Before we start

Eight cubes and twelve cubes – we do not know the size of each cube.

Questions

1) 500 cm³; 1500 cm³; 2000 cm³; 1300 cm³; 1400 cm³; 1000 cm³; 700 cm³

2) a) 6 cm³ 6 ml   b) 9 cm³ 9 ml
   c) 20 cm³ 20 ml   d) 36 cm³ 36 ml

Challenge

Amman has noticed that he can find the volume by multiplying the three sides so 2 × 3 × 2 = 12.

16 cm³ 16 ml;   18 cm³ 18 ml;   24 cm³ 24 ml

24 cm³ can be made up six ways:

1 × 1 × 24, 1 × 2 × 12, 1 × 3 × 8, 1 × 4 × 6, 2 × 2 × 6, 2 × 3 × 4

9.5 Imperial measurement (p.156)

Before we start

240 cm   3500 ml   4·5 kg

Questions

1) a) 90 cm   b) 3 m   c) 2·1 m   d) 15 cm

2) a) 84 lb   b) 126 lb   c) 280 lb
   d) 7 lb   e) 105 lb

Challenge

a) 320 pints   b) 6400 fluid ounces
a) 1 metre   b) 1 mile
c) 4 gallons   d) 8 stone

9.6 Calculating perimeter (p.158)

Before we start

54 m   58 m

Questions

1) Showers = 46·2 m; Changing area = 80·7 m;
   Children's pool = 74·7 m; Diving pool = 56·8 m;
   Main pool = 144·2 m

231

2) a) Estimates will vary and so therefore will the differences.
   b) Actual measurements are given here but may have small variations when measured.
   c) a) 21 cm   b) 14·4 cm   c) 18·4 cm
      d) 18·7 cm   e) 20.6 cm   f) 19.2 cm

Challenge

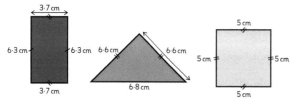

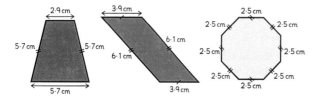

9.7 Calculating the area of regular shapes (p.162)

Before we start

Assuming whole numbers then 16 × 1, 8 × 2, 4 × 4

Questions

1) a) 36 m²   b) 78 m²   c) 192 m²
   d) 420 m²   e) 5000 m²

2) b) 10 cm²   c) 14 cm²   d) 12·5 cm²
   e) 9 cm²   f) 12 cm²

Challenge

Rectangle = 21 cm²;   Triangle = 12 cm²

Nuria could draw rectangles:

1 × 24, 2 × 12, 3 × 8 and 4 × 6

And triangles:

1 × 48, 2 × 24, 3 × 16, 4 × 12, 6 × 8

9.8 Converting metric units (p.166)

Before we start

First volume is 16 blocks = 1 × 1 × 16, 1 × 2 × 8, 1 × 4 × 4 or 2 × 2 × 4 (given)

Second volume is 18 blocks = 1 × 1 × 18, 1 × 2 × 9, 1 × 3 × 6 or 2 × 3 × 3 (given)

Questions

1) Amman = 80 cubic centimetres; Finlay = 120 cubic centimetres; Isla = 240 cubic centimetres; Nuria = 210 cubic centimetres

2) a) 40 cm²   b) 48 cm²   c) 54 cm²   d) 80 cm²

Challenge

Nuria's attempt = 32 cm²

Possible cuboids with volume 36 are 1 × 1 × 36, 1 × 2 × 18, 1 × 3 × 12, 1 × 4 × 9, 1 × 6 × 6, 2 × 2 × 9, 2 × 3 × 6, 3 × 3 × 4

Possible cuboids with volume 45 are 1 × 1 × 45, 1 × 3 × 15, 1 × 5 × 9, 3 × 3 × 5

10 Mathematics, its impact on the world, past, present and future

10.1 Mathematical inventions and different number systems (p.168)

Before we start

Answers will vary.

Questions

1)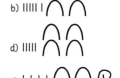

2) Answers will vary.

3) a)   b)
   c)   d)

Challenge

Answers will vary.

11 Patterns and relationships

11.1 Exploring and extending number sequences (p.170)

Before we start

1, 2, 4, 8, 16, 32, **64**. The rule is double the previous number.

Questions

1) a)

No. of sponges (s)	1	2	3	4	5
No. of candles (c)	3	6	9	**12**	**15**

   b) 3
   c) Number of candles = 3 × number of sponge cakes
   d) B = 3 × C   e) B = 27

2) a)

No. of cardigans (c)	1	2	3	4	5
No. of buttons (b)	5	10	15	**20**	**25**

   b) 5   c) Number of buttons = 5 × number of cardigans
   d) B = 5 × C   e) B = 40

Challenge

a)

No. of buses (b)	4	5	6	7	8
No. of pupils (p)	88	**110**	**132**	**154**	**176**

b) P = 22 × b

12 Expressions and equations

12.1 Solving equations using mathematical rules (p.172)

Before we start

a) 264   b) 107

a) =   b) =   c) <   d) <

Questions

1) a) 5   b) 4   c) 5
   d) 21   e) 4   f) 9

2) a) 4   b) 5   c) 20
   d) 16   e) 3   f) 2

3) Answers will vary.

Challenge

a) 146   b) 216   c) 194
d) 8   e) 44   f) 2

13 2D shapes and 3D objects

13.1 Naming and sorting polygons (p.175)

Before we start

Answers will vary.

Questions

1) a) regular octagon   b) irregular octagon

    c) parallelogram   d) equilateral triangle

2)

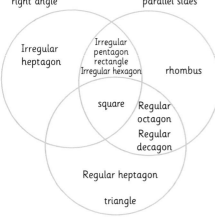

Has at least one right angle / Has at least one set of parallel sides / Is a regular polygon

- Irregular heptagon
- Irregular pentagon, rectangle, Irregular hexagon
- rhombus
- square
- Regular octagon
- Regular decagon
- Regular heptagon
- triangle

3) Answers may vary.

Similarities	Differences
1 Have straight sides. **2** No right angles.	**1** Decagon has 10 sides, triangle has only three. **2** Triangle has three vertices, decagon has 10. **3** Decagon has parallel sides, triangle does not.
1 Have straight sides. **2** Have parallel sides.	**1** Hexagon has six vertices, rectangle has four. **2** Rectangle has four sides, hexagon has six. **3** Rectangle has right angles, hexagon does not.
1 Have straight sides. **2** Have right angles.	**1** Pentagon has two right angles, square has four. **2** Square has four sides, pentagon has five. **3** Square is regular, pentagon is not.

Challenge
Answers will vary.

13.2 Describing and drawing circles (p.178)

Before we start
A and D are circles.

Questions

1) a)

Circle	Radius (mm)	Diameter (mm)	Estimation of circumference (mm)
A	15	30	90
B	7.5	15	45
C	12.5	25	75
D	9	18	54
E	6	12	36

   b) diameter is 2 × radius.

   c) see additional column in table above.

2) a)

Circle	Radius	Diameter	Circumference (approximately)
F	14 cm	28 cm	84 cm
G	25 mm	50 mm	150 mm
H	6 cm	12 cm	36 cm
I	4·5 cm	9 cm	27 cm
J	21 mm	42 mm	126 mm

b) Draw circles F to J using the info in the table.

3)

Challenge
Answers will vary.

13.3 Constructing 3D objects (p.180)

Before we start
The shape cannot be c).

Questions

2) b) i) square-based pyramid.

      ii) square-based pyramid, triangle-based pyramid.

      iii) cuboid. A cube also if we accept that a square is a special rectangle.

3) a) cube, cuboid.

   b) Cube and cuboid are less strong that the objects made with triangles.

   c) Triangles are the strongest shape. Objects made up of triangles are stronger and more stable than those made up of squares/rectangles.

Challenge
20   32   44

13.4 Constructing nets (p.182)

Before we start

3D object	Faces	Vertices	Edges
Cube	6	**8**	**12**
Cuboid	**6**	8	12
Triangle-based pyramid	**4**	**4**	**6**
Square-based pyramid	**5**	**5**	**8**

Questions

2) a) a – triangle-based pyramid

     b – cuboid

     c – cuboid

3) a, b, e, f, g

Challenge

14 Angles, symmetry and transformation

14.1 Identifying and sorting angles (p.184)

Before we start
c) is an obtuse angle, so it is in the wrong column. e) is a right angle and does not belong in either column.

Questions

1)

Acute	Obtuse	Reflex
a	c	b
f	i	d
j		e
		g
		h

2)

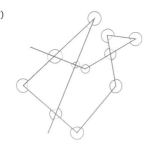

⌒ reflex	⌒ acute	⌒ obtuse
6	13	11

3) Has 4 or more obtuse angles   Has at least one reflex angle

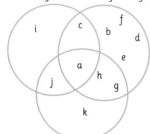

Has at least one right angle

Challenge

Amman is incorrect. An acute angle of over 45° and an obtuse angle do not necessarily add up to 180° or more.

This can be shown by drawing, or by calculating.

e.g. 50° (acute) + 110° (obtuse) = 160°, allowing for a reflex angle of 200°.

14.2 Measuring and drawing angles (p.187)

Before we start

You can't tell whether it is acute or reflex because the angle has not been marked. Depending where you measure it, both children are right.

Questions

1) a) 13°   b) 152°   c) 176°   d) 330°
   e) 81°   f) 236°

2) a) 330°   b) 82°   c) 312°   d) 61°
   e) 257°   f) 124°   g) 348°

3) Check drawings match the given angles.

Challenge

Answers will vary.

14.3 Finding missing angles (p.189)

Before we start

110° ± 2°

Questions

1) a) 35°   b) 54°   c) 63°   d) 9°   e) 74°

2) a) 105°   b) 156°   c) 47°   d) 94°   e) 147°

3) Allow some latitude in measuring.
   a) A = 55°   B = 35°   b) A = 145°   B = 35°
   c) A = 42°   B = 48°   d) A = 17°   B = 163°
   e) A = 67°   B = 113°

Challenge

a)   A is 39° and B is 51°   b)   C is 150° and D is 30°

c)   Answers will vary.

14.4 Locating objects using bearings (p.192)

Before we start

a)   B2   b)   1 square South East or 1SE

Questions

1) a) 045°   b) 225°   c) 135°   d) 315°

2)

From	Compass direction	Bearing	To
Circle	**East**	**090°**	Rhombus
Kite	North-east	**045°**	**Trapezium**
Rhombus	**North**	000°	**Square**
Trapezium	**South**	180°	Triangle
Square	**North-west**	315°	**Trapezium**
Star	**South-west**	**225°**	Pentagon

3) a) a) Cairn Toul   b) Beinn A Bhuird
       c) Ben Macdui   d) Braeriach

Challenge

Answer between 325° and 330°
Between 150° and 155°
040°

14.5 Reading coordinates (p.195)

Before we start

Shape is a parallelogram.

Questions

1) a) (1,7) (3,7) (3,4) (1, 4)
   b) (5,7) (6,7) (6,5) (7,5) (7,4) (5,4)
   c) (1,1) (2,3) (3,3) 4,1)
   d) (5,1) (6,3) (8,3) (7,1)

2) a) (0, 0) (0, 6) (5, 0) (5, 6)
   b) Answers will vary, but appropriate coordinates to make the letters M, Z and W.

3) a) (1, 6) (1, 7) (6, 7) (7, 6) (7, 5) (2,5)
   b) (2, 3) (7, 3) (7, 2) (6, 1) (1, 1) (1, 2)

Challenge

(5, 7) (5, 6) (6, 6) (6, 4) (5, 4) (5, 3) (3, 3) (3, 4) (2, 4) (2, 6) (3,6)

14.6 Line symmetry (p.197)

Before we start

Finlay is right. The shape is not symmetrical. It is not reflected in the mirror line.

Questions

1)

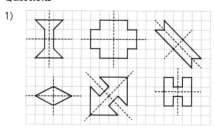

2)

No lines of symmetry	One line of symmetry	Two lines of symmetry
E, G, K	A, D, I	B, C, F, H, J

3)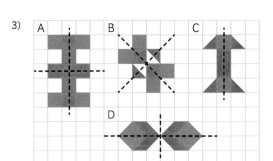

C is the odd one out as it has only one line of symmetry.

Challenge

B and D are correctly reflected in both lines of symmetry.

14.7 Symmetrical pictures and diagrams (p.200)

Before we start

The shape is not reflected in the mirror line. The shape is a parallelogram and therefore is not symmetrical.

Questions

1)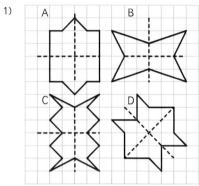

2)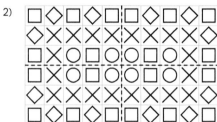

3) Check design of each square.

4) Answers will vary.

Challenge

Check designs of patterns.

14.8 Reading scale maps (p.203)

Before we start

c) 35 km

Questions

1) i) scale is 1 cm : 100 km. 8 cm on the map is 800 km in real life.

   ii) scale is 1 cm : 5 km. 8 cm on the map is 40 km in real life.

   iii) scale is 1 cm : 2 m. 8 cm on the map is 16 m.

   iv) scale is 1 cm : 150 km. 8 cm on the map is 1·2 km or 1200 m

2) a) 1 cm : 200 m

   b) 8 cm. 1600 m in real life, or 1.6 km

   c) 6 cm. 1·2 km in real life

   d) 15 cm

3) a) 1 cm : 50 cm

   b) The garden is approximately 7 m × 4 m to the nearest metre.

   c) i) 1·5 m   ii) 1·75 m

   d) 4 cm by 1 cm

Challenge

Answers will vary.

15 Data handling and analysis

15.1 Working with a range of graphs (p.206)

Before we start

They have collected time series data, so they should display it in a line graph.

Questions

1) a) 69   b) 2013, 2016, 2017, 2018

   c) 2013   d) 2016

2) a) 3500   b) 1000   c) February

   d) 20 500   e) 17 500

3) a) Erica   b) Charlie

   c) Bob, Charlie, Duncan   d) Bob, Charlie, Erica

Challenge

a) lorry B   b) lorry A   c) 11:30   d) 55 km

15.2 Using pie charts (p.209)

Before we start

a) is incorrect

Questions

1) a) Spain   b) 10   c) 6   d) 10 (15−5)

2) a) Woofit Mix   b) 13   c) 14   d) 13

3) a)

Hours spent watching TV on school days	Number of ten-year-olds	Angle in pie chart
0	7	**70°**
Less than an hour	11	**110°**
1–2 hours	13	**130°**
2–3 hours	4	**40°**
More than 3 hours	1	**10°**
Total	36	360°

b)

Hours spent watching TV

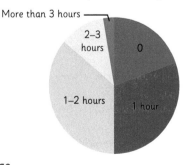

Challenge

a) 720   b) A   c) 320

15.3 Creating and interpreting graphs (p.212)

Before we start

Answers will vary.

Questions

1) a) bar graph   b) line graph
   c) pie chart is best but bar graph acceptable

2) a) True   b) True   c) False
   d) False   e) True

3) a) Additional statements will vary.
   b) i) True   ii) True   iii) True   iv) True

Challenge

Answers will vary.

15.4 Drawing conclusions from graphs (p.215)

Before we start

No scale, no title, axes are not labelled

Questions

1) a) Not appropriate.   b) Appropriate.
   c) Not appropriate since they may like football but it is not the favourite.
   d) Not appropriate.

2) Yes. This is what the data shows.
   Yes. This would be a possible extension to the investigation.
   No. The data doesn't show the water cooling.
   No. This isn't relevant to the temperature.

3) a) Not appropriate (see December).
   b) Appropriate.
   c) Appropriate – this would be an appropriate extension.
   d) Not appropriate.

Challenge

Answers will vary.

16 Ideas of chance and uncertainty

16.1 Investigating the possible outcomes of random events (p.218)

Before we start

Answers will vary.

Questions

1) a) Unlikely   b) Likely   c) Impossible
2) a) Unlikely   b) Likely   c) Impossible
3) a) Impossible   b) Likely   c) Unlikely

Challenge

Answers will vary.

© 2019 Leckie

001/30012019

10 9

The authors assert their moral right to be identified as the authors for this work.

ISBN 9780008313999

Published by
Leckie
An imprint of HarperCollins*Publishers*
Westerhill Road, Bishopbriggs, Glasgow, G64 2QT
T: 0844 576 8126   F: 0844 576 8131
leckiescotland@harpercollins.co.uk   www.leckiescotland.co.uk

HarperCollins Publishers
Macken House, 39/40 Mayor Street Upper, Dublin 1, D01 C9W8, Ireland

Publisher: Fiona McGlade
Managing editor: Craig Balfour
Project editors: Rachel Allegro and Alison James

Special thanks
Answer checking: Rodger Alderson
Copy editor: Louise Robb
Cover design: Ink Tank
Layout and illustration: Jouve
Proofreader: Dylan Hamilton

A CIP Catalogue record for this book is available from the British Library.

Acknowledgements

Images © Shutterstock.com